PENGUIN BOOKS

THE LAND WHERE LEMONS GROW

'A lovely exploration of citrus fruit in Italy'
Bee Wilson, *Sunday Times*, Books of the Year

'A fresh chapter was opened in Italian garden history
by *The Land Where Lemons Grow*, expertly unpeeling
Italy's citrus culture from Garda to Palermo'
Jonathan Keates, *The Times Literary Supplement*,
Books of the Year

'It would be a treat to find under the tree'
Carolyn Hart, *Daily Telegraph*, Books of the Year

'This is the first among my books of the year. Every page of
Attlee's subtle fusion of history and horticulture made me feel that
it's time to pack the bags again for Italy . . . this is very much a
boots-on-the-ground narrative, as she clambers into abandoned Tuscan
orangeries, decodes a Renaissance garden under a Genoese car park
and penetrates the sanctum of the Calabrian bergamot industry.
She sauces her erudite adventure with a scatter of
citrus recipes' Jonathan Keates, *Literary Review*

'Inspired and inspiring, in prose as sharp as the fruit it celebrates'
David Wheeler, editor of *Hortus*

'Fascinating, vivid, poignant, challenging, cheering . . .
A distinguished garden writer, Attlee fell under the spell of
citrus over ten years ago and the book, like the eleventh labour
of Hercules to steal the golden fruit of the Hesperides, is the result.
She writes with great lucidity, charm and gentle humour, and wears
her considerable learning lightly . . . Helena Attlee's elegant, absorbing
prose and sure-footed ability to combine the academic with the
anecdotal, make *The Land Where Lemons Grow* a welcome
addition to the library of citrologists and Italophiles alike'
Clarissa Hyman, *The Times Literary Supplement*

'Truly fascinating . . . For many years, Attlee has been collecting
evidence for a story of citrus trees in Italy. The result, *The Land
Where Lemons Grow*, is remarkable, excellently produced and
essential for all lovers of Italy, their summer libraries and
out-of-season itineraries . . . Attlee's book is . . . able for
anyone int   travel,
g   eat,

ABOUT THE AUTHOR

Helena Attlee is the author of four books about Italian gardens, and others on the cultural history of gardens around the world. Helena is a Fellow of the Royal Literary Fund and has worked in Italy for nearly thirty years.

HELENA ATTLEE

The Land Where Lemons Grow

The Story of Italy and Its Citrus Fruit

PENGUIN BOOKS

PENGUIN BOOKS

UK | USA | Canada | Ireland | Australia
India | New Zealand | South Africa

Penguin Books is part of the Penguin Random House group of companies
whose addresses can be found at global.penguinrandomhouse.com.

First published by Partiuclar Books 2014
Published in Penguin Books 2015
008

Copyright © Helena Attlee, 2014
Maps Copyright © Jeff Edwards

The moral right of the author has been asserted

Typeset by Jouve (UK), Milton Keynes
Printed in Great Britain by Clays Ltd, St Ives plc

A CIP catalogue record for this book is available from the British Library

ISBN: 978-0-241-95257-3

For Alex, of course

Contents

Contents

Acknowledgements

I began to gather material for this book long before I knew I would write it, and my thanks must go to all the citrus farmers, nursery-men and gardeners who have been so generously conversational over the years. I am especially grateful to Marco Aceto in Amalfi, Salvino Bonaccorso, Princess Maria Carla Borghese and Giuseppe Messina in Sicily, Paolo Galeotti, Gionata Giacomelli and Ivo Matteucci in Tuscany, Giuseppe Gandossi and Domenico Fava on Lake Garda, Pietro Donato, Antonio Miceli and his daughter Sara in Calabria and Danilo Pollero in Liguria.

I am much indebted to Professor Giuseppe Barbera for his kind-ness and enthusiasm, and for sharing his extensive knowledge of the political and economic history of citrus in Sicily, John Dickie for valuable insights into Mafia history, to Dr Chiara Nepi at the Nat-ural History Museum in Florence and also to Stena Paternò, Giorgio Galletti, Ezio Pizzi, Vittorio Caminiti, Franco Galiano and Schmuel Keller. My thanks to Dr Cathie Martin of the John Innes Centre in Norwich and Professor Paolo Rapisarda of CRA in Sicily for gener-ously sharing their research, and to Dr Justin Goodrich of Edinburgh University for patiently answering botanical questions.

I have travelled all over Italy in pursuit of citrus, journeys that would have been impossible without the kindness and hospital-ity of many people, including Margherita Bianca, Rudolf and Benedikta von Freyberg, Rachel Lamb, Marchese Giuseppe and his daughter Giulia Paternò Castello di San Giuliano and Cristina di Martino in Sicily, Lucia Rossi in Ivrea and Nick Dakin-Elliot in Tuscany.

As ever, I am grateful to my dear friends Valeria Grilli in Rome and Jenny Condie in Venice, whose enthusiasm and practical help have been essential at many stages of my research, to Alex Dufort for kindly testing the recipes in the book, to Sue MacDonald and Boxwood Tours for support in so many ways, and to Natoora (London) for generously supplying the fruit used for recipe testing.

My thanks also go to Anthony Ossa Richardson for his translations of Giovanni Battista Ferrari's *Hesperides* and to my agent, Antony Topping, of Greene and Heaton for good advice, humour and unstinting support. Thanks to all at Penguin, and in particular to my editor, Helen Conford, for her enthusiastic and incisive input.

Above all, thank you to Alex Ramsay and our three daughters for years of encouragement, and for their patience during my long absences, both in Italy and at my desk.

Citrus Crops in Italy

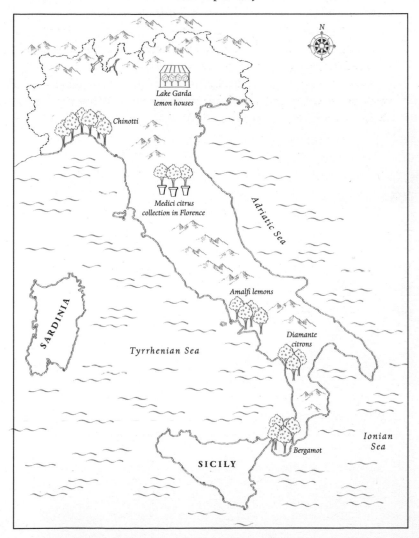

Citrus Crops in Sicily

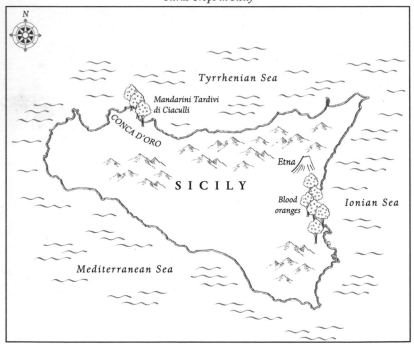

The Land Where Lemons Grow

The Scent of Lemons

I remember when planes were so expensive that people usually made the long journey from England to Italy by ferry and train. Once you got to Paris it was easy because you could catch the Palatino, a sleeper that sped you through the night towards Florence and Rome. I first made that journey over thirty-five years ago. At dawn I lifted a corner of the curtain in the stuffy couchette and realized we had already crossed the border. We were somewhere near Ventimiglia on the Italian Riviera, and there were lemons growing beside the station platform, their dark leaves and bright fruit set against a backdrop of nothing but sea. I never forgot those trees, or the way they charged the landscape around them, making it seem intensely foreign to my very English eye.

I didn't know it then, but travellers from northern Europe have always been thrilled by the sight of Italian citrus trees, and so my reaction was entirely predictable. Hans Christian Andersen, the Danish author and poet best known for his fairy tales, visited Italy in 1833, and when he saw citrus groves for the first time he responded with the mixture of rapture and envy that Italy can still provoke among visitors from colder and less romantic countries. 'Just imagine the beautiful ocean and entire forests with oranges and lemons,' he wrote to a friend; 'the ground was covered with them; mignonettes and gillyflowers grew like weeds. My God, my God! How unfairly we are treated in the north; here, here is Paradise.'[1] A sun-soaked, poetic image of Italy was particularly powerful in Britain after the First World War, when soldiers returned from the freezing trenches of Picardy and Flanders dreaming of the sensuous and hedonistic lifestyle they associated with the Mediterranean.[2] Captain Osbert Sitwell chose Sicily as the antidote to his bleak experience of war, a journey he describes in *Discursions on Travel*,

Art and Life, published in 1925. Here the orange serves as a symbol for all he loved about the Mediterranean. 'Where it grows,' he says, 'you will find the best climate, the most beautiful European buildings.'[3] And as his train trundled through orange groves outside Palermo, he remarked, 'About the whole tree there is a design, a balance, a geometrical intention and sense of grouping, an economy and right use of colour, that make it rank almost as high as a work of art.'[4] D. H. Lawrence, never a soldier, began the period of voluntary exile after the war that he referred to as his 'savage pilgrimage', a journey that took him to Sicily between 1920 and 1922. In 'Sun', a short story set in his sexually charged version of the Sicilian landscape, he returns again and again to images of citrus trees and their fruit, making Juliet, the angry and frustrated American heroine, meander naked through a 'dark underworld of lemons', discovering freedom and sensuality for the first time in her life.[5]

~

A few years after my first glimpse of lemons I returned to Italy as a student. I had chosen to live in Siena, and although Tuscan winters are much too harsh for citrus trees to grow outside all year, I got used to glimpsing pots of lemons in the sunny courtyards of city palaces and on terraces in front of country villas. When they disappeared from sight in winter I learned they had been taken into the shelter of purpose-built lemon houses, or *limonaie*. At first I assumed that Italians took their citrus trees for granted, rather as we do our apples in England. As my Italian improved, however, I began to realize that the trees and their fruit had a special place in the Italian imagination. When Galileo wrote *Dialogue Concerning the Two Chief World Systems*, the book that would lead to his conviction for heresy in 1632, he used oranges to illustrate the absurdity of the different values we give to the objects around us. 'What greater stupidity can be imagined than calling jewels, silver and gold "precious" and soil and earth "base"?' he wrote. 'People who do this ought to remember that if there were as great a scarcity of soil as of jewels or precious metals, there wouldn't be a prince who would not spend a bushel of diamonds and rubies and a cartload of gold just to have enough earth . . . to sow an orange seed and watch it sprout, grow and produce its handsome leaves, its fragrant flowers and fine fruit.'[6] With citrus on my mind, I returned to England for my final year at university. Here I fed my nostalgia for Italy with 'I Limoni', a poem published in 1925 by Eugenio Montale (one of the most important Italian poets of the twentieth century). Montale's lemons are not the romantic trees that provoked Hans Christian Andersen's outburst, and they do not grow in the sultry, sensual landscape of D. H. Lawrence's Sicily. Instead, they can be found on prosaic patches of rough ground at the end of rutted tracks, or beside miserable city streets in winter. And yet the perfume of their blossom (or *zagara* as it is in Italian) transforms even the bleakest and most banal landscape. It is both infinitely precious and freely given for anyone to enjoy. As Montale says, '*qui tocca anche a noi poveri la nostra parte di ricchezza/ ed è l'odore dei limoni*' – 'now it's our turn, us poor ones, to have a share of riches/ and it's the scent of lemons'.[7]

For many years my working life has been rooted in the elite world of Italy's gardens, both as a writer and as the leader of garden tours. This has made it easy to pursue the history of citrus as an ornamental garden tree, and yet as my interest grew, I realized that those pot-grown trees represented only a fragment of the whole story. During journeys that have taken me from the bergamot groves of Calabria, on the southern tip of the Italian peninsula, to lemon houses set against the snowy backdrop of the Alps, I found that citrus trees and their fruit have had a radical part to play in Italy's political and social history, and have brought extraordinary wealth to some of the poorest places in the country. Unlike those cosseted garden specimens, these trees grew in open ground, and like the oranges known as *wu nu*, or 'wooden slaves', in ancient China, 'they worked tirelessly to make and keep their families rich'.[8]

To find out about the working life of citrus I had to leave the comfortable territory of villa and palace gardens in Tuscany, Lazio and Umbria, get in among the trees of commercial citrus groves in southern Italy, and meet the men and women who spend their working lives there. I crossed the Strait of Messina to Sicily, where the best blood oranges in the world grow in the shadow of Mount Etna on the eastern side of the island. Travelling west, I found orange, lemon and mandarin trees in a strange liminal landscape between Palermo, the mountains and the sea.

Many citrus farms in Sicily and southern Italy are in remote and deeply rural landscapes, where foreign visitors are rare and only dialect is spoken. I soon discovered that a penknife came in handy in the orange groves of those places, because most fruit clings to the tree and, unless you cut the stem from the twig, you risk tearing its skin. I also learned that an orange is never peeled in the field. There's a ritual to observe, and that's another reason for an orange farmer to carry a penknife. First he holds the fruit in the palm of one hand, stem side up. Then he makes a horizontal cut to divide it neatly in half. The juice of a fresh orange is lavish, uncontrollable, and scent explodes into the air. He tosses the top half into the long grass because, in the orange, both juice and sweetness are concentrated in the bottom, furthest from the stem. Then he cuts a slice and offers it on the flat side of the knife's blade. I've taken part in this tiny ritual in fields all over Italy, but it's always a strangely touching moment, and I enjoy its intimacy just as I used to love it when someone lit a cigarette for me. The flavour of an orange straight from the tree is incomparable.

Curious Fruit

Citrus collectors in Renaissance Tuscany

⟡

I t was the sheltered lives of elegant pot-grown trees that taught me my first real lessons about citrus. The process began in 1987, when I wrote my first book on Italian gardens, illustrated with photographs by Alex Ramsay.[1] We travelled the length and breadth of the country with our baby daughter, seeing gardens all day and sleeping, a little, in a tent at night. Vast terracotta pots containing ancient citrus trees lined the paths of almost every garden we visited. It was late summer when we began work and, although the oranges on the trees were still unripe, there was plenty of fruit of other kinds. We were almost always alone, but there is a code of honour associated with garden visiting and we didn't eat it. In late August we ignored warm figs on trees in Lazio, in September we averted our eyes from laden vines smothering pergolas in Tuscany, and as the summer slipped into a long, golden autumn, we turned our backs on ripe pomegranates and glowing persimmons in the walled gardens of Venice. My resolve didn't weaken until the end of November, when we found ourselves in a semi-abandoned garden beside the Brenta Canal in the Veneto. An overgrown path lined with pots of citrus and flowering geraniums ran the width of a garden below a castle, and it was an orange fallen from a pot-grown tree that finally seduced me. I picked it up, tore off the brittle peel and crammed a segment into my mouth. The dry seedy flesh was so sour it seemed to cauterize my tongue. These days I can recognize a bitter Seville orange at a glance, but I knew nothing then and I'd been given the only lesson an ignorant orange thief could understand.

A few years later I began to take groups of English visitors to the gardens we had explored. The gardeners seemed to disappear as soon

5

as we arrived, although I'd sometimes glimpse the front wheel of a barrow protruding from a row of citrus pots, or see the top of a battered trilby hat moving between their branches, and then I'd pursue them. It was those men, and in my memory they are invariably old and male, who gave me my first lessons in the art of cultivating citrus. Many of them had inherited jobs from their fathers, along with expertise and intimate knowledge of each individual tree in the garden. They introduced me to a specialist language, to their own techniques for repotting, feeding and treating diseases. 'A lemon is like a man when it's ill,' said Silvano Mazzetti, a gardener who had long since retired but seemed unable to stop coming in to work every day at Villa Poggio Torselli in Tuscany. 'The iller it is, the pricklier it gets, just like a man who gives up shaving because he feels poorly.'

I began to scrutinize the citrus trees in every garden I visited as carefully as pictures in an exhibition. Some seemed capable of bearing two or even three kinds of fruit at the same time, and the fruit came in an extraordinary variety of sizes, shapes and colours. The oddest and most arresting were made like yellow hands, with an unsettling number of shiny fingers that were sometimes clasped together, as if in prayer, and sometimes splayed, like a beggar's hand with spiders instead of coins nestling in its palm. Others had rotund bodies attached to protuberances like long quizzical noses, so they resembled a cluster of watchful creatures on the branch of the tree. In one place I saw fruit patterned with orange and green stripes, and in another I discovered a citron, a monstrous yellow fruit as big as six large lemons bound together by deeply folded, lumpy skin. And at around this time I began to read about the fashion for collecting rare or exotic plants and trees in Renaissance and baroque Italy, and I realized that many of the trees producing these strange fruits, with bizarre shapes, ribbed or pitted skins, warts and carbuncles, were the fragments of great citrus collections that had belonged to the richest and most important families of the sixteenth, seventeenth and eighteenth centuries. Knowledge of this historical role is essential to any understanding of the weight and volume of Italy's relationship with citrus fruit.

During the seventeenth century the villa and palace gardens that housed Italy's citrus collections became part of a wider intellectual landscape. They were outdoor extensions of the collections of curiosities, or private museums, made up of objects from distant places and times that were assembled by educated, affluent or aristocratic gentlemen all over Europe. Citrus, with its enormous variety of fruit and curious mutations, was an important element in many of these collections. English philosopher and statesman Francis Bacon gave a vivid impression of the kind of collection that might be found in the house of a learned gentleman anywhere in Europe. Indoors there would be '. . . a goodly, huge cabinet, wherein whatsoever the hand of man by exquisite art or engine has made rare in stuff, form or motion; whatsoever singularity, chance, and the shuffle of things hath produced; whatsoever Nature has wrought in things that want life and may be kept; shall be sorted and included'. The setting for a citrus collection was described as 'a wonderful garden, wherein whatsoever plant the sun of divers climate, or the earth out of divers moulds, either wild or by the culture of man brought forth, may be . . . set and cherished'. The garden also housed rare birds, animals and fish, so that it represented '. . . in small compass a model of the universal Nature made private'.[2]

During spring and summer citrus trees lined the paths and steps of the garden and were arranged in groups around fountains and statues. In winter their pots huddled in the shelter of the *limonaia*, or lemon house. Such close proximity allowed them to cross-pollinate freely, so that each collection continued to grow in size and diversity. They evolved like a series of ancient and incestuous families inhabiting the walled gardens of villas and palaces all over Italy, where they existed like members of enclosed communities, outliving countless generations of aristocratic owners and expert gardeners. Renaissance and baroque gardens were full of entertainments, such as complicated mazes, dripping grottoes and elaborate tree houses. Water was everywhere; it sprang in unexpected and powerful jets from the façades of buildings, from the cracks between flagstones and the seats of garden benches. It powered hydraulic instruments and moved statues so they seemed to shift of their own accord. There were exotic creatures in the menagerie, rare birds in the aviary and brightly coloured fish in the garden pools. The complicated botany of citrus made it sit well among these wonders, ensuring that there was always something mysterious about it, something fluid and unknowable, so that its fruit was endlessly surprising and entertaining.

The study of botany developed rapidly at this time, but there were many aspects of citrus that botanists and citrus collectors did not understand. Sudden changes in temperature, periods of drought, unusually high rainfall or even wind could trigger mutation. This often affects only one or two branches of a tree, which flower at a slightly different time, produce fruit that matures at a different rate, or is even an entirely different shape and colour from the fruit on the rest of the tree. When this happened to a tree in a citrus collection it was a wonderful mystery and, at this time of scientific exploration, it made citrus an even more fascinating and desirable collector's item.

Botanists and collectors would also scan the trees for the early signs of lemons with the strange, grasping fingers of the kind that I first saw in a Tuscan garden. Like tulip collectors waiting for valuable 'breaking' bulbs in seventeenth-century Holland, they had no idea what caused their lemon trees to produce such odd fruit. It wasn't until the twentieth century that the real reason for fingered lemons was finally discovered. The culprit was found to be a microscopic mite, *Aceria sheldoni* (citrus bud mite), that attacks the buds of lemon blossom, causing the damaged flowers to produce peculiar digitated fruit. The mite is a pest, but its impact is still considered so delightful that it is known as *acaro delle meraviglie*, or 'mite of the marvels'. Collectors in the sixteenth and seventeenth centuries also took great delight in another fingered fruit, a variety of citron we now call *Citrus medica* var. *sarcodactylis*, from the Greek *sarkos*, meaning 'fleshy', and *dactylos*, meaning 'finger'. The tree is native to China and north-east India and it produces fingered fruit naturally, without the interference of the bud mite. The fingers are created when the carpels forming the interior of the fruit fail to fuse together very early in their development, and remain distinct from one another. They are made entirely from pith covered in a highly scented rind. In China, where it has been cultivated since the tenth century, it is called 'Buddha's Hand', displayed as an ornament and used to scent rooms and clothes.

~

Digitated lemons and other fruit with strange deformities have always been known as *bizzarrie* or 'bizarres' in Italy. The finer the citrus collection, the more *bizzarrie* it contained (although the unpredictable aspects of citrus biology would always make any collection intrinsically unstable). The Medici family's collection in Florence was indisputably the finest in Europe. Its history dates back to 1537, when Cosimo I de' Medici (1519–74) came to power and inherited Castello, the family's rural retreat on the edge of the city. You hardly need a map to find your way from the centre of Florence to Castello today. It's enough to head north towards hills that appear from time to time between *palazzi* and high-rise blocks, arriving at last to squeeze the nose of the car into a shady space between two trees, and step out into the tranquil air that still clings to villa and garden.

These days you can see from one end of Castello's garden to the other, and yet it wasn't always so. The garden that Niccolò dei Pericoli, sculptor, garden architect and engineer, designed for Cosimo incorporated many degrees of shade. Giorgio Vasari writes about Pericoli and the garden at Castello in that vast compendium of biographical and artistic detail *Le vite dei più eccellenti pittori, scultori e architetti*, published in 1550. He explains that Pericoli was always called Il Tribolo, or 'trouble', a nickname given to him as a child because he was '. . . so high spirited in every action, that he was always cramped for room and was a very devil both among other boys at school and everywhere else, always teasing and tormenting himself and others, so that he lost his own name of Niccolò and acquired that of Tribolo to such purpose that he was called that ever afterwards by everyone'.[3] Tribolo laid out the new garden in 1539, complicating the open site by dividing it in two with a wall and turning the larger, lower garden into a network of open-air rooms set out around a central maze and linked by shady pergolas and carefully controlled vistas that made it impossible to see the whole site at once.

During the sixteenth century members of the educated elite associated gardens with the classical Golden Age, and in particular with the eleventh of the twelve life-threatening, muscle-building, sweat-inducing, intellectually challenging labours of Hercules, when he was required to steal golden apples from the Gardens of the Hesperides.[4] The Medici used the iconography of statues and fountains to make a connection between their family and the heroic virtue, determination and strength of Hercules, and so it was natural to find a fountain at the centre of Castello's new garden, surmounted by the figure of Hercules wrestling with Antaeus. Ever since the second century AD, artists had represented the golden apples of the Hesperides as different varieties of citrus fruit, and Tribolo strengthened the link between the Medici and Hercules by turning the new garden into a citrus paradise.

The walls of the lower garden were soon swathed in citrus cordons, so that the French naturalist Pierre Belon, who visited Italy between 1546 and 1549, could already describe oranges and lemons covering them 'like a tapestry'.[5] Montaigne, the French essayist and traveller, thought the villa at Castello 'not worth looking at' when he visited in 1581, but he was delighted by the garden. 'In every direction, you see a variety of arbours,' he reported, 'thickly formed of every description of odoriferous trees, cedars, cypresses, orange trees, lemon trees, and olive trees, the branches of which are so closely interwoven that the sun, at its meridian height, cannot penetrate them.'[6]

Giorgio Vasari described the upper garden at Castello as an area dedicated entirely to citrus. Today the narrow space is sheltered only to the north by the base of a hill, but originally the other sides were also enclosed by walls. Cosimo's trees were the common crop plants of Sicily and the south, but when his eldest son, Francesco I de' Medici (1541–87), took over, the collection became very much more varied. In 1585 Agostino del Riccio, author of several treatises on gardens and natural history, described eleven rare citrus trees and their wonderfully curious fruit.[7] Among them were several oranges. One from Palermo was so spectacularly sweet that you could bite into it like an apple. Others provoked an added frisson by being 'pregnant'. The skin of a pregnant fruit burst open to reveal a uterus packed with tiny babies, and del Riccio remarked that some even had protuberances resembling the fingers of a tiny hand emerging from their centres. He also mentions the flavour and relative juiciness of many of the fruits in the collection, although remarkable *bizzarrie* like these were rarely picked or put to practical use. Instead, they stayed on the trees, to be admired like pictures in an exhibition.[8]

~

By the end of the sixteenth century the Medici collection had overflowed into Boboli, the gardens behind the Pitti Palace in Florence. It continued to expand throughout the seventeenth and early eighteenth centuries, but when Gian Gastone de' Medici died in 1737, his sister Anna Maria Luisa bequeathed all the family's property – its villas and palaces; libraries full of books and manuscripts; collections of paintings and antiquities; scientific instruments and natural curiosities; statues; furniture and jewels amassed by generations of her ancestors – to Francis I, Duke of Lorraine. She made the bequest on the condition that nothing should be removed from Florence, and this embargo covered the vast and multi-generational collection of citrus trees that had lived their entire lives in heavy terracotta pots, like a tribe of babies who never left their prams. Francis I may not have appreciated the Medici citrus legacy, but his son Pietro Leopoldo, who succeeded him as Duke of Tuscany in 1765, was profoundly influenced by the Enlightenment and this made him passionately interested in the natural sciences, and also gave him an understanding of the enormous importance of preserving such an ancient and varied botanical collection. While it belonged to the Medici, the collection was appreciated for its ornamental qualities, as a source of private entertainment and the focus of amateur scientific enquiry. When it passed into Pietro Leopoldo's hands it took on a new role as a tiny element among the social and educational reforms he introduced to Tuscany.

When he arrived in Florence, Pietro Leopoldo found the majority of the people shockingly uneducated, their lives blighted by famine and the inefficient bureaucracy, high taxes and old-fashioned legal system imposed by the Medici and left unreformed by his father.[9] He modelled his new administration on Enlightenment principles, taking it upon himself to disseminate information about science and the natural world in the belief that knowledge was a tool that could be used to combat suffering, superstition and tyranny. He believed in giving the citizens of Florence a chance to educate themselves, and to this end he founded a new science and natural history museum called La Specola in Via Romana, the narrow street that runs from Palazzo Pitti to the Porta Romana. When its doors opened in 1775, La Specola was the first museum in the world to be accessible to the general public, although initially a distinction was made between the lower classes, who could enter between eight and ten in the morning, 'if decently dressed', and 'the intelligent and well educated', who had free access from one o'clock, as long as they removed swords and overcoats and left them by the door. Felice Fontana, the museum's first director, summarized its ethos when he explained that it was opened 'to enlighten people and make them happy by making them civilized'.[10]

Over time, Pietro Leopoldo amassed all the scientific and natural history collections gathered by the Medici under one roof at La Specola, including such treasures as Galileo's own scientific instruments and new specimens that Felice Fontana sought out for him all over Europe. Of course, the Medici citrus collection could not be brought indoors, but this problem was resolved when accurate, life-size, scientific wax models and plaster casts were made of all the citrus fruit that grew in the Boboli Gardens. The models fulfilled several different purposes. They found their place among the exhibits in the museum, they were used for teaching students of botany, for instructing farmers in the lucrative practice of pomology and for recording and commemorating particularly successful grafts, or fruit from trees in danger of extinction. There is no record of the exact date of their creation, although the models are known to have been made in the *officina di ceroplastica*, a wax workshop on the ground floor of La Specola. Here the modellers produced life-size versions of several different kinds of fruit and vegetables, as well as exotic plants brought back from the New World and corpses. Corpses? Yes, the bodies and body parts of paupers and unclaimed patients were carried in wicker baskets across the city from the hospital of Santa Maria Novella. In the *officina* they were dissected and used for casting accurate models destined for use as teaching aids upstairs, in the school of anatomy.

La Specola is still a public museum, although you'll find only the anatomical waxes there because these days the natural history museum is divided, and the fruit and flowers are displayed in the botanical section, in Via La Pira. Nevertheless, it's well worth going to La Specola, because it's only by seeing the full range of models made in the workshops that you begin to grasp the extraordinary skill of the wax modellers. The museum is on the third floor, and if you happen to arrive between lectures you will be pushed and jostled by students hurrying up and down the broad steps that have been worn away by the feet of generations of their predecessors, exposing a layer of fossils embedded in the stone. The last time I went there I found myself in a room full of wax models of horrifying obstetric and gynaecological incidents, with only a small bespectacled French boy for company. Nothing in that room was all right and I watched anxiously as he studied each obstetric mishap long and hard. I wished my French were good enough to say something light-hearted or comforting, and soon he dashed away, his urgent footsteps diminishing rapidly down the corridor. Moments later a rush of thundering steps brought him back with two slightly larger sisters. More earnest gazing, more silence, and then the three of them bolted off together and returned dragging their father. Good luck, I thought, as they began to voice their anxious questions.

The wax workers' *officina* must have been a busy place, cluttered with cauldrons full of melting beeswax, marble slabs for rolling it out, modelling tools, weighing scales and the jars of vegetable dyes, pigments, resin and linseed oil that were used for conditioning and colouring the wax. Plaster fruit was made from casts in much the same way as the wax. Plaster enabled the modellers to reproduce every nuance of the fruit's form and the subtly changing textures of its skin. When plaster fruit emerged from the cast it was white, but layer after layer of pigment was applied to reproduce the exact colour of the original.

There was a long tradition in Italy of making fruit for ornamental purposes and yet nothing like Pietro Leopoldo's scientifically accurate casts had ever been created in Europe before. Handsome cedar-wood cases were specially made for storing and displaying them, and for many years they were shown in the museum alongside a series of citrus paintings made by Bartolomeo Bimbi for Cosimo III de' Medici. The citrus canvases were part of a much larger commission that saw Bimbi painting every kind of fruit to be found growing in Tuscany. Once again, these decorative objects served several purposes. Many of the paintings recorded rare or monstrous animals, exceptionally large or strangely shaped vegetables and peculiar flowers, anomalies that might easily be forgotten and never seen again. Bimbi's commission also served to glorify the rich and fruitful territory of the Grand Duchy. He started work in 1699, and by the time he completed the citrus paintings in 1715, he had already produced meticulous and clearly labelled portraits of figs and apples, pears, plums, cherries, peaches, apricots and grapes. All of them were destined to hang at Villa Topaia, a country villa a few miles north of the city, that Cosimo used as a retreat from public life. The four citrus paintings hung in an antechamber next to his bedroom.[11] Each one was over two metres wide, and together they portrayed 116 different varieties of citrus fruit. Bimbi painted with oil on canvas, rendering each fruit with photographic accuracy at life size.

Today you must go to the still-life museum in the Medici villa at Poggio a Caiano to see the original canvases, but they are often reproduced in prints. Bimbi gave the fruit a regal air by spreading it across a trellis supported by elaborately carved, classical herms. A shield at the bottom of each canvas was inscribed with the names of the exhibits. Perfectly ordinary fruits nestled alongside teratological, or monstrous, specimens of gargantuan proportions or bizarre shapes.

Cosimo III's daughter, Anna Maria Luisa, commissioned Bimbi to produce three more paintings after her father's death in 1723. I think these close studies of exceptionally large lemon–citron hybrids (or 'citrated' lemons) are the most beautiful citrus paintings Bimbi ever produced. In two of them the vibrant yellow fruit is displayed on blue and white china plates against a luxurious backdrop of red drapery, its bumpy skin rendered with the loving attention due to any life model.

The citrus casts and waxes and Bimbi's paintings continued to be displayed at La Specola until the mid-nineteenth century, but when photographic techniques that could be used to record botanical specimens began to develop, the importance of the citrus exhibits diminished, and eventually the fruit and its display cabinets were removed from the exhibition hall to make way for more contemporary material.

~

It was a sunny morning in early May when I first heard about the casts made from fruit in the Medici collection. I had arrived in Pisa the previous night and needed to return to the airport to receive a group of garden visitors at about lunchtime. The rest of the week would be spent showing them gardens in Tuscany, but until then I was free. I came out of my *pensione* into warm spring sunshine and headed for the nearest bar. When I'd ordered a cappuccino, I grabbed the local paper and found a table in the sun. The centre spread was all about citrus. There was an article about a citrus nursery near Lucca, publicity for a citrus show in a garden nearby and a long piece about the Medici citrus collection.

By now I could think only of citrus, but I had an appointment to visit the herbarium at the botanic garden and, although I was in no mood for examining dried plant specimens, I hurried over there as soon as I had finished reading. I gazed glumly at specimens of seaweed mounted on card in the eighteenth century and longed to be outside, and preferably among citrus trees.

My guide was a professor of botany, and before long I clumsily brought the conversation around to citrus. She seemed to understand: she was an academic, after all, and subject to her own botanical passions. She showed me up to the library on the first floor and suggested I take the opportunity to consult the citrus section. The room was bright with spring light and a strong breeze blew through the open windows, lifting the pale curtains and the pages of the first book I took off the shelf. It was a history of orchards and ornamental fruit in Tuscany written by Mariachiara Pozzana and published in 1990.[12] In it I read about a series of plaster casts made during the eighteenth century from fruit in the Medici citrus collection. They had recently been rediscovered, and Pozzana suggested that once these strictly accurate casts were restored and identified it would be possible to rebuild the collection, returning it to the glory of the late eighteenth century. By now I was running late, but as I left for the airport I bumped into the professor again in the corridor. When I stopped to thank her, I mentioned the casts. She knew all about them and scribbled a couple of names and phone numbers on to a scrap of paper. 'I don't know where they are now,' she said, 'but one of these people will be able to tell you.'

My chance encounter with the professor, whose name I never knew, was the beginning of a quest that would become the undercurrent of my life for almost a decade. I kept the scrap of paper she had given me for four years before I managed to make some free time in Florence and arrange to see the casts. Even then, I didn't have long, because I was on my way to a conference in Lucca, but it was the best I could do. One of the telephone numbers I had been given belonged to Dr Chiara Nepi, head curator at the botany department of the Museo di Storia Naturale, part of the natural history museum, in Florence. Chiara was immediately and unquestioningly helpful. She knew that the casts were being stored at an office in the Boboli Gardens and she offered to make an appointment for me.

~

I arrived in Florence in autumn clothes on an unexpectedly hot October day, uncomfortable and prosaic among women still dressed for summer. My appointment at Boboli was for 2.30 p.m., time enough for lunch, and I found myself in Piazza Santo Spirito, where the sun blazed on the fallen leaves of plane trees that should have been shading the square from its heat. I sat at a table under a canvas awning and ordered *risotto al radicchio*. It came cupped in a purple chicory leaf, the glistening grains of rice a pale pink. When I'd finished eating I pulled out a copy of Chiara's email confirming my meeting. 'Ring this number ten minutes before you arrive,' she said, 'and then go straight to the amphitheatre in the Boboli Gardens.'

This was the first of many emails I was to receive over the years, emails as cryptic as clues on an extended citrus treasure hunt. The first two steps were to be repeated again and again – arrive at a certain location and make a call to an unknown recipient. In Florence I had to follow up by going through the dark heart of the Pitti Palace to the amphitheatre in the Boboli Gardens immediately behind it. I could have done with a third instruction, something telling me what to do when I got there. Instead, I arrived in the gardens, looked for clues in the faces of Japanese tourists and watched two old women making slow progress across the front of the amphitheatre in the shade of black umbrellas. What was I looking for? I had no idea. Eventually I approached a guard in uniform. 'Ah, the signora who wants to talk about lemons,' he said at once, and led me across shimmering expanses of hot gravel to the gardens' administrative offices.

When I arrived, the atmosphere was grumpy. It seemed that I had already put everyone to immense trouble, because the rediscovered casts had been lost again and it had demanded great dedication to find them. This news did nothing to dampen my own excitement. It was a long time since the casts had been rediscovered and I was anticipating shining citrus sculptures, a radiant celebration of the glorious diversity of citrus. I was delighted by their quasi-magical ability to get lost, be rediscovered and lost again, and now I had finally tracked them down, I imagined I could hold a cast in my hand and feel a direct connection with the Medici collection at the end of the eighteenth century.

As soon as I saw the box I remembered that research trips rarely turn out as expected. It was made from cardboard and it had started life as a container for bags of gnocchi, and then got very damp. Inside, the fragile casts were casually swaddled in rags, old vests and dusty carrier bags. There were twenty-three fruits and I lifted each one carefully from the box, laid it on the table and removed every layer of its dusty wrapping. It was like unwrapping Christmas decorations, although the contents of each parcel were far from festive. Some of the paint clung to the surface of the casts, a ghost of yellow, a hint of orange, but most had disintegrated and turned to the colour of mould, so that they looked more like gourds left too long in a bowl than citrus fruit. Only their plaster shapes and the realistic textures of their skins were unchanged.

It was clear that in the gardens of the Medici the lemon had been an infinitely protean thing. Its yellow skin might have been patterned with broad green stripes or split to reveal a second fruit nestling like a Russian doll inside it. Its shape was equally inconsistent and I found a double-breasted lemon, a pair of Siamese lemons, joined at the stem, their surfaces painstakingly stippled to give the impression of the skin's open pores, and lemons with horns. The other fruit in the collection had been subject to equally fascinating and peculiar mutations, and in the depths of the box a huge citron cowered, its side smashed to reveal plaster-of-Paris guts. The essential oils in the skins of real citrus fruits make them shine, but the fruit on the table had matt, dusty pelts. They looked like a gathering of sad, citrus ghosts. Eight of the casts had been restored and labelled, and they were just as colourful as I had expected. Someone had wrapped them in tissue paper and put them in a wooden fruit tray, as if restoration gave them the rights of real fruit.

It had no curator, that fallen fruit, but every so often the grumpy man who had shown me into the office returned to see how I was getting on. He seemed unable to believe that I was really interested in the citrus casts and assumed that I, like him, would find more amusement in the antics of a blind cat outside the window. 'Look at this!' he'd say, beckoning me over, and we would lean on the windowsill, elbow to elbow, looking long and hard at an unremarkable cat in the gardens far below. 'Did you come to look or to talk?' he asked eventually. It was a good question. 'I had hoped to talk to someone,' I said, suddenly feeling the bathos of the occasion. That's when he told me about Ivo Matteucci, Boboli's head gardener, and 'the only person here who really knows anything about the casts'. Ivo finished work at 2 p.m. 'Why didn't you come this morning?' grumpy man asked, blithely unaware of the tortuous arrangements I had made to arrive exactly when I was told. We had to wait for Ivo to ride his Vespa all the way home to San Casciano, some fifteen kilometres south of Florence, pick up our message and come bumping all the way back. It was worth the wait. Over the years I've met many head gardeners in Italy and, when he burst in, I wasn't surprised by his energy and friendly enthusiasm. Suddenly I wasn't the only mourner at the grave. Ivo told me how he first found the casts at the back of a damp cupboard in his office, but even he could not explain why their restoration had been abandoned.

~

My next appointment was with Dr Chiara Nepi, who had offered to show me the wax models made in the *officina di ceroplastica* and originally displayed alongside the plaster casts at La Specola. By the time I left Boboli I was already late and so I flagged down a cab. The air inside was thick with smoke and incense from a joss stick that was burning on the dashboard and turned the small space around it sacred. We were silent, the driver and I, as we crawled through the rush-hour traffic. Eventually we arrived in Via La Pira, a narrow street behind Piazza San Marco, and I found myself feeling refreshed, as if I had just spent fifteen minutes in a Japanese Zen temple instead of locked in Florence's rush hour.

Although the taxi left me at the door of the museum, I spent some time searching for the botany department, which I eventually found behind an anonymous white door on the top floor of the building. And there was Chiara Nepi. The sight of her white lab coat made me realize that I had left the dangerously haphazard world of Boboli's citrus casts and entered a safer, more controlled environment. Nothing could be broken or lost here, where Chiara was in charge of high rooms lined with glass-fronted cupboards and copies of Bartolomeo Bimbi's vivid eighteenth-century paintings of outlandish fruit and enormous vegetables.

The wax fruit collection couldn't have been more different from the dusty orphans I'd seen at Boboli. I'd left the orphanage now and arrived at the heart of a thriving family made up of waxes modelled from the citrus collection, and from other kinds of fruit, vegetables and exotic flowering plants imported into Tuscany from all over the world. The models of oranges, lemons and citrons were also made in the *officina di ceroplastica* from plaster casts taken from real fruit in the Boboli Gardens. As soon as the cast set, the inside of it had to be rubbed with soft soap to block any pores in the plaster before the wax could be strained into it. The wax was then built up inside the cast in subtly coloured layers. Once it was cool, the fruits could be turned out, polished and finished, if necessary, with additional touches of colour on their skin.

The produce of this complex process was arranged across six shelves in a glass-fronted cabinet. Each fruit was displayed, just as it always had been, on a handsome gilded-wood platform looking something like a cake stand. The wax fruit must always have been superior to its plaster counterpart and it was often sold to private collectors who put it to purely decorative use. Unlike the plaster casts I'd seen at Boboli, the wax models had the luminosity of real citrus fruit. It would be wrong to call them 'beautiful', because that's not the word for a warty and bifurcated *Limone scanellato di fior doppio o scherzoso* (a 'grooved, double-flowered or joking lemon'), a name as long as the unending title of an Italian aristocrat, or *Limon sponginus*, an appropriately open-pored and spongy-looking fruit so broad and carbuncled that it resembled a toad, or even the very ordinary northern Italian *Limon S. Remo*. Nevertheless, all of them had been restored, and though they may not have been beautiful, they were certainly magnificent.[13]

TAGLIOLINI ALLE SCORZETTE
DI ARANCIA E LIMONE

When I ate this pasta and tasted its surprising orange and lemon sauce for the first time, I was careful to ask which kinds of lemons were used in the recipe. I was in Settignano, a village high above Florence, where Damiano Miniera had founded a dynamic restaurant and wine bar called Enoteca la Sosta del Rossellino. Damiano was from Sicily, where lemons are a major crop, and yet he insisted that any variety of lemon would do for his *tagliolini*. He wasn't jealous of his recipes and he would recite this one to anybody who could hear it on the crowded restaurant floor.

2 oranges
1 lemon
a knob of butter
¼ onion, chopped
a big splash of white wine
100ml single cream
salt and freshly ground black pepper

- Peel the fruit, being careful to exclude the pith. Cut the peel into razor-thin slivers and cook them in boiling water for 5 minutes or so to remove some of the bitterness, then drain them.
- Melt the butter in a small frying pan and add the onion. When it's translucent, pour in the white wine. Add the drained peel, together with the juice of the oranges and the lemon and the cream.
- Simmer for 5 minutes before seasoning with salt and black pepper and sloshing the sauce over a bowl of warm pasta.

I couldn't know it at the time, but there was no need to rely on the citrus casts and waxes to make that direct physical connection with the original Medici citrus collection. All I had to do was return to the garden of Castello, although it has changed enormously since the Medici owned it. The new eighteenth-century rulers of Tuscany didn't appreciate the old-fashioned complexity of Tribolo's Renaissance design. They swept away the maze and moved the fountain at its centre, pulled out the shady scented tunnels and uprooted parterre beds and exotic trees, flooding the garden with light and exposing it to view from end to end. There's very little shade now and in midsummer the temperature on the enclosed south-facing slope regularly climbs to forty degrees Celsius. That's too hot for lemons, which are at their best between fifteen and thirty degrees; when the temperature rises, many of them stop growing and drop their flowers. Nevertheless, if you visit today you will find about a thousand pots of citrus lining the paths.

The curator of this magnificent collection is Paolo Galeotti. These days he is a handsome man in his early fifties, but when he first came to work in the garden at Castello he had only just left agricultural college in Florence. Citrus can't be grown as a commercial crop in Tuscany and consequently he'd been taught almost nothing about oranges, lemons or mandarins. Fortunately he knew just enough to recognize that some of the citrus trees in the garden were very old, and he began to find out as much as he could about their history. This was the beginning of an enormous research project that took him to national libraries, state archives, historic gardens and ancient citrus groves all over the country, and eventually made him a leading authority on the history of Italian citrus and the co-author of a book on the subject.[14] He has spent thirty years developing practical skills to meet the most intimate needs of the ancient and precious trees in his care. This gives him the advantage of a close, first-hand knowledge of all the trees and fruit he writes about.

Among the pots lining the paths at Castello are some ancient, distorted, asymmetric, unlovely trees that are at least 300 years old and would already have been part of the Medici collection when Pietro Leopoldo inherited it in 1765. They stand out from all the rest, making the garden strangely reminiscent of an old people's home. Instead of armchairs lining the walls there are pots, and instead of faces marked by a lifetime of experience there are blackened, split branches and twisted trunks emerging from knots of gnarled rootstock. Paolo Galeotti has spent years piecing together the events that gave the trees their extraordinary appearance. According to his research, their regular, well-managed existence continued undisturbed until the First World War, when the *limonaie* that gave them shelter each winter were converted into a military hospital and the tender, cosseted, ancient trees were left outside in the cold to fend for themselves.

Citrus has always seemed to welcome grafting, pruning and other forms of human interference, and both ornamental and commercially cultivated trees are generally made up of two parts that have been grafted together. In most cases the upper tree, or scion, is one species of citrus and the lower trunk and roots, known as the rootstock, is another, chosen for its hardiness and resistance to disease. The survival of the trees at Castello can be explained by the fact they were grafted on to sour orange rootstock because it is the only kind of citrus hardy enough to survive the cold.

The two parts are grafted by inserting a bud and a sliver of bark from the branch of the fruit-bearing tree into a T-shaped cut in the trunk of the seedling used as rootstock. The two are then bound tightly together and left in peace to heal their wounds and find a common purpose. This technique has always enabled citrus farmers and gardeners to combine the most vigorous and disease-resistant root system below ground with the best fruit-bearing tree above ground. Sour orange has always been a popular rootstock because of its hardiness and although many of the trees at Castello died back when they were left outside over winter, their roots survived and sprouted new growth from below the graft, reverting to being sour oranges through and through.

When the Second World War broke out, citrus trees came very low on the list of things to protect and the Castello citrus inventory, meticulously preserved and updated for many centuries, was mislaid. Many of the trees lost the lead labels inscribed with their inventory numbers, and some were even chopped up and burned as firewood.

As Galeotti discovered more about the original Medici citrus collection, however, he began to restore some of the diversity that had been lost by grafting new material on to the sour orange rootstock of the surviving trees. It is this combination of ancient, knotted rootstock and new growth from the grafted material that gives the oldest trees at Castello their extraordinary appearance. Galeotti is particularly pleased by a tree that combines branches grown from the original sour orange rootstock with newly grafted material from a lemon, so that it bears two different kinds of fruit. He admits that visitors to the garden sometimes complain about the ugliness of the trees, but he always defends them by saying, 'We don't get rid of our old relations just because their legs don't work and they can't run.'

In the course of his research Galeotti read about the *bizzarria*, the most famous, highly prized and altogether extraordinary specimen in the Medici collection. It was a chimera, an example of the strange fruit that can occasionally be produced when two different citrus species or varieties are grafted together and the tissues from the rootstock combine in the graft with those of the scion. In Greek mythology the chimera was a fire-breathing monster with the head of a lion, the body of a goat and the tail of a snake or a dragon emerging from its back. Chimeric fruit displays a similar jumble of features and is generally found on the branch immediately above the graft.

The first written record of the chimera known as the *bizzarria* was published in 1644 by Pietro Nati, director of the botanic garden in Pisa, who called it *De malo limonia citrata aurantia vulgo la Bizzarria*. This is an earnestly descriptive title for a fruit that displayed the characteristics of both the sour orange and citron–lemon cross known as a citrated lemon. Its skin was patterned with green and orange stripes, and the pale yellow flesh beneath the skin was that of the citrated lemon.

The *bizzarria* was discovered for the first time in the garden of Villa Torre degli Agli, now uncomfortably close to Florence airport, where the road name, Via del Giardino della Bizzarria, commemorates its existence. At first the gardener claimed to have created it himself, but according to Paolo Galeotti, 'when they got him up against the wall, he admitted to having stumbled upon the *bizzarria* by chance in the garden'. Needless to say, this fabulously strange and exotic fruit found its way into the Medici collection, and for a century at least it was only to be seen in gardens belonging to the family. However, when Tuscany and the entire Medici estate passed to the House of Lorraine, the *bizzarria* was released and it began to crop up all over Europe. There are records of it growing among the trees of Europe's other great citrus collections at Versailles, in Potsdam and in the botanical garden of Amsterdam, where J. Commelyn created the finest citrus collection in northern Europe. Then the trail goes cold and from the mid-nineteenth century the *bizzarria* seems to disappear without trace, although Darwin refers to it in *The Origin of Species* (1859).

It was not until 1980 that Paolo Galeotti noticed a sour orange tree at Castello with a single twig bearing leaves quite different from those on the rest of the tree. He suspected at once that he might be witnessing the reappearance of the *bizzarria*. Bravely, he cut the twig from the tree and grafted it on to sour orange rootstock, 'and three years later,' he told me, 'the grafted twig bore fruit, and I realized immediately that I had rediscovered the famous *bizzarria*'. You may be lucky enough to see the *bizzarria* if you visit. It looks a small and inconsequential tree with some leaves that are twisted, wrinkled and striped with darker or lighter shades of green. Just like the original chimera, it combines the physical characteristics of a sour orange with those of a citron and a lemon. Galeotti has ensured that the *bizzarria* will never disappear again by creating other specimens through grafting and taking them to the Boboli Gardens and the Orto Botanico in Florence, where they can still be seen.

~

With Galeotti in charge, life for the citrus trees at Castello is much the same as it was under the Medici. All of them grow in the terracotta pots known as *conche* in Tuscany. The pots in the Medici family's gardens have always been handmade from clay found in the soil at Impruneta, a village near Florence that has been famous for manufacturing terracotta for hundreds of years. These highly porous containers create ideal conditions for citrus as it can't tolerate being waterlogged. Galeotti only has to rap his knuckles against a pot to know if a tree needs watering. A dry pot makes a long, echoing sound, like a bell.

Each pot is numbered with the year the tree was repotted. Most of the trees are not permitted to grow very large, and although Galeotti never cuts the roots, he will grate the ends off a few of them before choosing a new pot of exactly the right size. Modern citrus nurseries tend to force their plants into rapid growth by overfeeding them. This makes the trees look wonderfully lush when they are sold, but such a system produces weak, tender trees that are prone to disease. At Castello they use only organic fertilizers and the trees are pruned once a year, using traditional principles. Galeotti gives the oranges a compact shape and thins out the branches of the lemons to make the trees hollow at the centre. There are no examples of the nineteenth-century goblet-shaped form known as the *coppa Toscana*, which you find in gardens all over Italy. The goblet shape is created by growing the trees' branches around a circular framework and leaving the centre of the plant hollow. Galeotti considers the reintroduction of this 'mistaken technique' as a cynical move by nurserymen, who can use a single branch curled around a frame to create the appearance of a tree. To his eye – and mine – a traditional, pot-grown tree with its short, sturdy trunk and the curved embrace of its branches is a much more beautiful thing.

Cooking for the Pope

Oranges, lemons and citrons were widely used in the kitchen in the sixteenth and seventeenth centuries. In 1565 Agostino Gallo, an author from Brescia in northern Italy, remarked, 'Everybody knows that much is to be gained from every kind of citrus tree . . . a good amount of money can be made from many parts of the plants.' He went on to give examples of the many uses known for citrus of all kinds:

> . . . citron blossoms can be eaten in salads or preserved in vinegar to be served with apples or sugar; even orange blossoms can be used for making superb scented waters . . . unripe fruit is made into many delicate condiments, and tiny oranges used for wreaths that are beautiful to look at and delightful to smell. The ripe fruit is valued for eating, making preserves and is given to the sick and also used in medicines . . . even the peels are sold for making good pickled fruit relish, orange juice, focaccia, liqueurs and spiced bread.[1]

Perhaps the most important collection of recipes at this time was a 900-page treatise published under the title *Opera* in Venice in 1570. Considered the culinary *summa* of its day, its author was Bartolomeo Scappi, who had worked in the households of a series of Italian cardinals before becoming private cook to Pope Pius V.[2] One of the six books in Scappi's *Opera* is dedicated to the preparation of every kind of meat, including porcupine, bear, dormouse, cows' udders, calves' eyes and stags' testicles. This was food for the elite, food designed to impress or even intimidate the guests gathered around the table.

As a modern reader, one is immediately struck by the ubiquitous presence of sour oranges among these recipes. *Citrus aurantium*, the sour orange, was the first kind of orange to arrive in Italy. It is the progeny of cross-pollination between two of the three most ancient species of citrus, pomelo (*Citrus maxima*) and mandarin (*Citrus reticulata*). The sour orange is an equal cross between the two, whereas the sweet orange, which reached Italy much later, is a back-cross, a second-generation hybrid with more mandarin than pomelo in its make-up. These days *melangoli, melaranci or cedrangoli,* as they are known in different parts of Italy, are used mostly to make marmalade or candied peel, and yet for 200 years at least the unique, earthy and aromatic taste of sour orange juice was as essential to an Italian banquet as ketchup is to fast food. Take Scappi's tortoise pie, for example. The pungent, spicy flavour of sour orange juice was essential to the success of a dish best made in autumn, when tortoises are at their finest. You begin by cutting off the creature's head, although elsewhere Scappi warns that a tortoise's body can live for a day without it.

TORTOISE PIE

Take a tortoise – which is much better than the sea turtle because it has more juice – cut off its feet and its head, remove it from its shell, skin it and clean it, removing its nails and all other foulnesses, including the bile from the liver. Cover all over with cinnamon, cloves, nutmeg and salt, a little sugar, the juice of several sour oranges and some saffron. Put in a pie dish, adding mint and onions fried in a little butter or oil. Cover in pastry and cook. Just before it is ready, season with sugar, verjuice and a little more sour orange juice. Complete the cooking.

Next, Scappi might mix sour orange juice with oil for steaming truffles, or stir it into the rich juices in a roasting pan and combine it with a little grape-must syrup, vinegar and a clove of garlic to make gravy. He often sprinkled sour orange juice over a dish at the end of its long and complex preparation. Take his recipe for the milk-filled udder of a cow. Having sealed the teats, he wrapped it in a piglet's caul and parboiled it in water with some ham. And then he un-wrapped it, covered it in pork lardons basted with pepper, cloves, cinnamon, salt, nutmeg and fennel, and roasted it on a spit. And when it was done, he sprinkled it with breadcrumbs and served it in slices with plenty of sour orange juice.

～

The juice of sour oranges was often used to lubricate a pungent mixture of sugar and precious spices, such as cinnamon, cloves, nut-meg and saffron, that had drenched the meat and fish eaten at banqueting tables ever since the Middle Ages. It is traditional to assume that these powerful combinations of flavours were used to disguise the rancid taste of rotten flesh, but that's a modern assump-tion. Food historians point out that when spices began to be used in the Middle Ages, animals were generally slaughtered close to home, and their meat was either consumed immediately or salted. It is more likely that spices, with and without sour orange juice, were added almost indiscriminately to medieval and Renaissance recipes because they were thought to play an important role in regulating the digestion and balancing the four bodily humours.[3]

The doctrine of humours first emerged in ancient Greece and was thoroughly explored by Galen (AD 131–200), whose texts were enshrined in Italian university syllabuses throughout the Renais-sance. By then it was common knowledge that the humours (blood, yellow bile, black bile and phlegm) determined a person's health, that it was important to keep them in balance and that each humour had a certain temperament or personality. A person with humours out of balance might be lucky enough to become 'sanguine', although they were equally likely to be 'choleric', 'melancholic' or 'phlegmatic'. Fortunately, the humours could be adjusted and brought back into kilter through diet. Each humour corresponded to one of four basic qualities: heat, cold, moistness and dryness. Food was also categorized as hot, cold, moist or dry, so that it could be used to temper the bodily humours, bringing them back into a healthy equilibrium. Spices were generally categorized as hot and dry, and this made them the ideal antidote to foods classified as wet and cold. Spices were combined with sugar to counteract the over-whelming saltiness of preserved meat or fish, and with sour orange juice to add another layer of flavour and lubricate the mixture.

Sour oranges also appeared at banquets as decorations, their dark-gold skins blazing bright against the virginal white of the table-cloth, undiminished by the rich colours of tapestries and brocades on the walls, and chiming with the golden thread woven through carpets underfoot. And neither were they lost in the extraordinary display of valuable objects on the banqueting table. They shone in the candlelight, their reflections glowing on the surfaces of gold and silver plates, and in the shining curves of crystal glasses and delicate decanters. They were often arranged in unpeeled slices around the edge of a serving dish, their edges crimped into a decorative frill, impaled on a spit in alternating patterns with songbirds and saus-ages, or simply piled in glowing heaps among an abundance of oysters. Candied oranges were always present, whole or in slivers among other kinds of candied citrus peel served at the end of a meal to aid digestion, or finely chopped and combined with sour orange juice to make a sauce for veal.

～

The subjective nature of taste has made it very difficult to know exactly when the first sweet orange, *Citrus sinensis*, reached Italy. Crusaders returning from the Near East in the twelfth and thir-teenth centuries brought a slightly sweeter orange called *Citrus aurantium* var. *bigaradia* home with them. Its flesh is not as bitter as the true sour orange, and although it was celebrated for its slightly less bitter flavour, it is not genuinely sweet. The Portuguese are generally credited with bringing the first truly sweet oranges back from India when Vasco da Gama discovered the sea route around the Cape of Good Hope in 1498, and for the next 400 years oranges were known all over Italy as *portogalli*, or different dialect versions of that word, and they were called 'portugals' in every other European language. Nevertheless, when Father Alvaro Semmedo, head of the Jesuit order in China, wrote a letter in 1640 about oranges in Canton, there could be no doubt that they were infinitely superior to the Indian oranges already familiar in Europe. In Semmedo's opinion, 'The oranges of Canton might well be queens over our own, in fact some people hold that they are not so much oranges as muscat grapes disguised.'[4]

The first of these truly magnificent sweet oranges arrived in Italy from China only in the mid-seventeenth century. Their arrival coincided with the advent of a new, simpler style of cooking in Europe. The pungent cocktail of spices, sugar and juice so prevalent in Scappi was gradually losing popularity. In the Middle Ages it had been prestigious to cook with spices because they were extraordinarily expensive, but as trade routes between Europe and the Far East opened up, their value gradually diminished, and heavily spiced dishes lubricated with the juice of sour oranges lost their cachet, although the fashion for spices endured in Italy long after France and England had adopted a simpler and more delicately flavoured style of cooking.[5] A German botanist visiting the country at the beginning of the eighteenth century was still able to observe that 'the Italians do not value sweet oranges half as much as the sour ones, the latter being used by them in connection with every kind of roast meat, the juices being pressed over it; likewise are these same sour oranges ... regarded by them as more palatable even than lemons'.[6] However, he was witnessing the end of the sour orange's supremacy. Its popularity had been due in part to the fact that many of the lemons grown in southern Italy and Sicily were of the mild or even semi-sweet variety known as *lumia*. As different varieties of lemons became available, many proved reliably bitter and cooks no longer had to rely on the acidity of sour oranges.

TORTA CAPRESE ALL'ARANCIA

There aren't many Italian recipes that use sour oranges these days, but this chocolate tart from Capri is a wonderful example of the kick that a few drops of the essential oil extracted from their skins can give. I made it for the first time when my friend Valeria arrived for the weekend with a large bag of fresh almonds from Puglia ('so much better than American almonds', she assured me) and a tiny bottle of tawny bitter oil.

200g dark chocolate (70%)
200g slightly salted butter
250g whole peeled almonds
200g caster sugar
5 eggs
¼ teaspoon drops essential oil of sour oranges
icing sugar
1 sweet orange

- Preheat the oven to 200°C/160°C fan oven, gas mark 6.
- Break the chocolate into squares and cut the butter into cubes. Melt them together in the top of a double boiler, stirring with a wooden spoon, and remove from the heat.
- Blitz the almonds in a food process to give a mix of 'flour' and chunks, then combine them with the sugar.
- Separate the eggs. Beat the yolks until they are light and foaming, then add the almonds and sugar, the melted chocolate and the essential oil. Beat the egg whites until they hold their shape, then fold them into the mixture.
- Tip into a loose-bottomed cake tin 26cm in diameter and 4cm deep that you have lined with greaseproof paper. Bake for approximately 40 minutes. Serve cold with a dusting of icing sugar and fine strips of sweet orange peel scattered over the top.

The decline of the sour orange as a culinary ingredient had begun, but the tree had a different future. It is a tall, vigorous specimen, a lovely tree with dark leaves that act as a foil to fruit of a deeper, darker colour than most other oranges. Sour orange trees are robust by nature. They have a strong root system, they adapt readily to a range of different soils and tolerate neglect. They are disease-resistant, can survive eight degrees of frost for short periods and, given ideal conditions, they can live for six centuries or more. Ever since the nineteenth century these characteristics have made sour orange popular both as an ornamental tree and as a choice of rootstock for more fragile varieties of citrus. Disturb the soil below a sweet orange, a lemon or mandarin and you'll often notice the air beneath the tree filling suddenly with the aromatic and very specific scent of sour oranges.

Golden Apples

A case of taxonomic havoc

⸺⁀◌⁀⸻

I t comes as no surprise after visiting Castello to hear that taxonomists have always struggled to pin down the complexities of the
citrus family and come up with a foolproof system of classification. Originally citrus was a very compact clan, a small and widely
scattered genus made up of only the mandarin (*Citrus reticulata*),
native to China, the pomelo (*Citrus maxima*) that grew in Malaysia
and the Malay archipelago, and the citron (*Citrus medica*) growing
on the slopes of the Himalayas in northern India. Wherever these
three ancestral trees were brought into contact with each other by
human migration or trade, they cross-pollinated. Most plants can
only cross-pollinate successfully with other plants of their own species, but citrus is unusual because cross-pollination between different
species is generally successful and produces viable seed. Many of the
familiar fruits cultivated today are the hybrid progeny of spontaneous
cross-pollination between wild or cultivated citrus trees. For example,
oranges (both sweet and sour) are hybrids between mandarin and
pomelo, grapefruit is the result of a pomelo–orange cross and the
lemon is a hybrid between citron and sour orange. Viable new fruit is
sometimes produced as the result of mutation, and a desirable
mutation can be perpetuated through grafting. Over the centuries
hybridization and mutation have been the source of an ever-expanding
number of varieties of familiar citrus fruit – the orange alone accounts
for 4,000. These tendencies have made the citrus family such a complex and unstable thing that David Mabberley, a contemporary expert
in citrus taxonomy, refers to 'taxonomic havoc'.[1]

Early attempts to resolve the confusion surrounding citrus
taxonomy were made among the great citrus collections of

seventeenth-century Italy, and one of the first people to attempt this daunting task was Giovanni Battista Ferrari, a Jesuit priest and professor of Hebrew at Collegio Romano, the Jesuit seminary in Rome. His book, *Hesperides, sive, De Malorum aureorum cultura et usu*, 'Hesperides, or, The Cultivation and Use of Golden Apples', published in 1646, was a determined attempt to create a taxonomy that would account for all the intermediate fruits: the hybrids and chimeras. It was the product of a scientific revolution that overturned the accepted wisdom of the classical age and transformed the study of natural history. Empirical research was replacing old-fashioned dependence on ancient texts, and by the time Ferrari was writing, natural scientists were fixated on the need to make an accurate and complete visual record of the natural world, and to create classification systems that would encompass both the native flora of Italy and the ever-increasing flow of exotic species from the New World.

Ferrari was not content with the subjective guesswork that formed the basis of earlier citrus classification systems and he underpinned his research with close, detailed observations of each fruit, counting segments and seeds, sampling juice and recording the colour, texture and thickness of skin. These details were accompanied by information on the name given to the fruit in different parts of the country and on its culinary and medicinal uses. He gathered this information by devising a citrus questionnaire for circulation among citrus growers all over the Italian peninsula. The questions encompassed a variety of subjects, including the fruit's name, the origin of the name, the appearance of the tree, its leaves, flowers and fruit, and the uses of the fruit, the propagation and cultivation of the trees, the diseases they suffered from and how to cure them. He gave responsibility for circulating the questionnaires to his friend Cassiano dal Pozzo, one of the most important aristocratic patrons and collectors in Rome. Dal Pozzo was well connected and he sent Ferrari's questions to an enormous variety of recipients, including dukes, cardinals, farmers and gardeners, who responded with alacrity, as if they too were longing for clarification of the complex, confusing citrus family tree.[2]

Cassiano dal Pozzo was extraordinarily important in the realization of Ferrari's great project. He assisted with every stage of the research, from circulating the questionnaires to collating the responses to them, although he was already engaged in a pressing project of his own, the creation of a *museo cartaceo*, or paper museum. Unlike a conventional seventeenth-century museum or collection of curiosities, which would display real fossils, stuffed birds and animals, shells and antiquities, dal Pozzo's paper museum was a collection of over 7,000 watercolours, engravings and prints recording all these things on paper. Among them was a large collection of exquisitely accurate coloured drawings by Vincenzo Leonardi celebrating the extraordinary diversity of citrus fruit. Cassiano generously allowed Ferrari to use a number of these drawings as the basis for engravings, for which he paid Cornelis Bloemaert, a Dutch painter and engraver working in Rome. Citrus was a particularly challenging subject for a taxonomy because every citrus-growing area in Italy nurtured its own varieties and devised its own set of common names. This created a general state of confusion and made even the most basic conversation about citrus difficult. Take Nicolas Peiresc, the French humanist, antiquarian and citrus collector living in Aix-en-Provence. When questioned, Peiresc confessed he couldn't tell the difference between a lemon known as an Adam's Apple and another called the Apple of Paradise, or the local sweet lemon and sweet lemons from Spain.[3] His problems – and he had many more – were a vivid example of the confusion surrounding the classification of citrus that Ferrari was attempting to resolve by creating a modern, accurate and clear-cut taxonomy. He might have been overwhelmed by the sheer promiscuity of citrus, but with the humour typical of his writing, Ferrari simply compared the difficulty of creating a stable taxonomy to the challenge facing Hercules when he had to steal the golden apples from the Gardens of the Hesperides. 'Since it is a labour worthy of Hercules,' he said, 'I think it is an excellent subject for me as well; and, as he exhausted himself in performing this labour, I am wearing myself out in writing about it.'[4]

Ferrari divided citrus into three rigid categories: citrons, lemons and oranges. He was fearless about classifying the hybrid forms that were the cause of such taxonomical confusion, and he created a whole category for teratological, or monstrous, fruit, which he gave the charming title of *frutte che scherzano*, or 'joking fruit'. His nomenclature system was highly descriptive and almost painfully detailed, so that names such as 'the distorted, digitated Adam's Apple lemon' (*Limon Pomum Adami distortum et digitatum*) became official titles that continued to be used long after Linnaeus invented his international system of plant nomenclature in 1749.[5]

Ferrari did his best to give scientific explanations for the odd shapes of teratological fruit. He said pregnant oranges were a product of the extreme fertility of the soil. He noted that lemons often looked like fingers and toes, but could equally well swell into two breasts, or resemble the horns of cows or deer, or even birds' beaks. In his opinion these bizarre mutations were caused by a partial abortion of the fruit on account of weak seeds. 'Generally,' he remarked, 'we shudder at deformity in humans or animals, but we love them in fruit.'

Ferrari had grown up with a medieval system of scientific thought that he could fall back on when he failed to find the answers he needed in modern, evidence-based research. He also resorted to unashamedly fanciful explanations for the peculiar morphology of certain fruit. Thus the indentations at the base of the Adam's Apple were the 'perpetual marks of Adam's teeth', and the strange shape of *Citrus medica* var. *sarcodactylis*, the Buddha's Hand citron, was explained in one of a series of extraordinary myths reminiscent of Ovid's *Metamorphoses*. All of these stories were illustrated with engravings made by Cornelis Bloemaert, from the drawings of Nicolas Poussin; Andrea Sacchi; Pietro da Cortona; Francesco Albani and Domenichino, some of the finest artists in Rome. The fingered citron triggers a myth about Harmonillus, a young man whose sweet singing voice so infuriates a witch that she transforms him into a citron tree. In this case Bloemaert's engraving, taken from a drawing by Sacchi, shows the shocking metamorphosis from man to tree, capturing the moment of fusion, when his legs take root and his hands turn into grotesque fingered citrons.

Most of the botanical plates in *Hesperides* showed two views of each fruit. In one the fruit was whole and in the other it was cut in two to reveal pith, segments and seeds. The leaves, twigs, flowers and buds were also rendered with an accuracy, sophistication and beauty unique at that time, and finally the name of each specimen was inscribed on an elegant flowing banner. David Freedberg, Professor of the History of Art at Columbia University and an expert on Ferrari, describes *Hesperides* as a watershed in the history of botanical illustration: 'Never before had the insides of citrus fruit been depicted with such care, never had the surfaces of their peel been shown with such obsessive attention to every kind of texture, rugosity, lump and protuberance.' You can experience the shocking excellence of Bloemaert's work by visiting the British Library in London and requesting *Hesperides*. Inside its scuffed leather binding the coarse pitted paper is saturated with ink. To see the engravings is to feel the weight of the fruit in your hand, the texture of its skin against your fingers and, if you are lucky, to experience something of the passion it aroused among citrus collectors.

~

Taxonomy forms the core of *Hesperides*, but Ferrari overlaid it with a colourful mosaic of anecdotes, recipes, medicinal remedies and customs. It is this element of his work that reveals the extraordinary number of uses for citrus in seventeenth-century Italy. He describes football tournaments in Reggio Calabria, on the toe of the peninsula, where oranges were used '. . . instead of balls in trivial and foolish play'. The fruit was prepared by boiling, its pulp removed and replaced with flax or weeds to make it lighter. Finally a twig was attached to each orange as a handle, to make it easier to throw, and 'as they fly in sport, gold flashes through the air'. Ferrari gave detailed descriptions of *limonaie* behind the high walls of gardens belonging to Rome's grandest families, the best recipe for pot-pourri and 'a million medicinal uses for the citron', including this medicine to be taken before breakfast to make the heart lighter, strengthen the stomach and improve the breath. This translation is a slightly condensed version of the original.

A CITRON TONIC

Take six citrons. Quickly remove the skin down to the flesh. If some skin sticks to the flesh, it doesn't matter, but may in fact be helpful. Put the skins into a very large spherical glass vessel. Pour in three pounds of spirit distilled from wine. Carefully stop up the vessel. Heat and preserve the same vessel in the sun for twelve days. After those days take the same spirit, now tinted with the colour of the citron, and smelling and tasting of citron, and strain it into another glass vessel, then add to it a very finely chopped candied citron. Now open the vessel and allow it to sit for twenty days in the sun. Then, having passed the liquid through a sieve, store in another well-stoppered vessel and take three drams of the mixture in the morning, an hour before breakfast.

A Day in Amalfi

I once arrived at dusk in Sorrento, on the steep shores of the Gulf of Naples. It was December and the trees in the piazza were strung with Christmas lights. At first I thought I saw golden decorations among their branches, then I realized they were the enormous rough-skinned lemon–citron hybrids that grow on the coasts of Campania. Sorrento and Amalfi lemons are the most famous local fruits, and they are the main ingredients in myriad local recipes that depend on their wonderful and very specific qualities, and are almost unknown outside a restricted local area. Their juice has very low acidity; you wouldn't expect it to give a kick to home-made mayonnaise, and you can't use it to replace the vinegar in a salad dressing because it is altogether too gentle for jobs like that. Instead, a summer pasta dish can be made by dressing spaghetti with raw garlic, parsley and the mild but intensely flavoured juice of a Sorrento lemon, and in Amalfi a slice of lemon sprinkled with ground coffee and sugar serves as a simple *digestivo* after a meal. In the sixteenth and seventeenth centuries the juice of Amalfi lemons was mixed with honey or sugar to make a sherbet. When the Viceroy of Naples came to town for an Easter banquet in 1602 he was served a sorbet made from lemon sherbet mixed with snow. The snow had been conserved in *nivare*, purpose-built ditches dug into the mountainside, high above the town. These days Amalfi lemons are more often peeled, thinly sliced and combined with garlic, olive oil, a drop of white wine vinegar, chopped mint and salt. In my experience, eating raw lemons is always an endurance test, but for locals the bright taste of these sparkling yellow salads is the very essence of summer.

Limone femminello sfusato amalfitano have been cultivated ever since the twelfth century. *Fuso* means 'spindle', and the *sfusato*

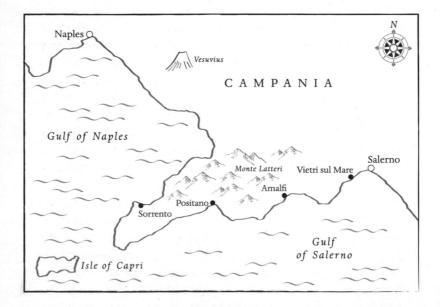

amalfitano has a slightly elongated, spindle-like shape. It's a large lemon with juicy flesh and thick skin full of highly perfumed essential oils. Inside, there are never more than nine segments and very few seeds. These unique qualities have been recognized by the European Community, and it has received the IGP (Indicazione Geografica Protetta), which they say is 'reserved for quality products which are guaranteed authentic only if they are grown in a very specific, limited area'. This is an accurate description of the Amalfi lemon, whose territory extends along the coast of the Bay of Salerno to either side of Amalfi, from Positano to Vietri sul Mare. The trees grow on broad terraces known as *macerine*, a name specific to Amalfi, though citrus terraces elsewhere are called *terrazzi* rather than *terrazze*, the Italian for terraces in any other context. Amalfi lemon trees are always grown inside a supportive scaffolding of slender chestnut posts. The posts are cut from the chestnut forests on the sides of the Monti Latteri, mountains named for the milk-giving flocks that graze them. Towering sheaves of chestnut poles can always be seen piled beside the mountain road high above the town. In winter, the lemon terraces are protected from heavy rainfall and hail by a covering of canes or netting supported by these posts. The harvest begins each year on 1 February, when the fruit ripens in lemon gardens closest to the sea. Further inland, where the climate is cooler and damper, the last lemons are picked at the end of October. As the summer progresses the flavour of the fruit gradually intensifies.

Growing Amalfi lemons has always been a labour-intensive business. Chestnut poles, stone for building the retaining walls of terraces, manure and even topsoil have to be laboriously carried up the hillsides from the valley floor. At harvest time, sacks of lemons weighing fifty-seven kilos each must be carried down the hill. Tending to the trees is a job for a gymnast happy to clamber about on the terraces and negotiate steep flights of narrow steps while carrying heavy equipment. The armature of chestnut poles must be replaced at regular intervals, although some farmers have begun to substitute metal for wood. This alters the timeless appearance of the lemon groves and demands a substantial investment, but once it's in place, the metal will last for much longer. Traditionally lemon gardens have been handed down from father to son, but by the mid-twentieth century low profits and hard manual labour held diminishing appeal, and gradually terraces that had been cultivated for centuries began to be abandoned.

The main street in Amalfi leads out of town to the entrance of the Valle dei Mulini (the Valley of the Mills), where the steep slopes are clothed in vineyards and lemon terraces. Today the sides of the valley are an intricate mosaic of brown and green. There is still more green than brown, but each brown patch contains an empty house and an abandoned lemon garden, the dead trees baked crisp by drought and years of relentless sunshine. This loss spells danger for the coast, where the steep cliffs are prone to landslides, and the stone terraces and the roots of the trees have stabilized the ground above for many centuries. Without them, there are fears of major landslides similar to the deadly disaster that killed over 160 inhabitants of the nearby town of Sarno in 1998.

Some families refuse to be pessimistic about the future of lemon farming. They have joined forces to set up the Cooperativa Amalfitana Trasformazione Agrumi, a farmers' cooperative focused on protecting cultivation and bringing young people back into the business. Marco Aceto, a founder member of the cooperative, offers tours of the enclosed, ancient world of his family's lemon gardens, where narrow stone steps link the terraces and grass grows lush in the velvet shade of the trees. When the tour is over, Marco's mother suddenly appears beneath the trees to offer us thick slices of lemon cake, sweet lemon salad and tiny glasses of home-made limoncello that we raise to the future of the *sfusato amalfitano*.

～

One of the most important cookery books of the eighteenth century is Vincenzo Corrado's *Il Cuoco galante*. In it he includes this recipe for citron and lemon sorbet.

Take three pounds of sugar and heat until it dissolves. Add the peel of a citron and the juice of several lemons until you achieve the flavour you require. Add two glasses of water and stir thoroughly. Sieve through a linen cloth into a watertight, tubular pot. This quantity of sorbet mixture will require eight parts of ice to one and a half of salt. The container must be closed and surrounded by ice that you have chopped and mixed with salt, so that you can turn the container easily and continuously. When you see that it has frozen to the depth of a finger, break it up and repeat the process until the sorbet is pale and soft. Serve in earthenware bowls.

In Amalfi in August, at the height of the holiday season, I turned my back on the crowded beach, pushed through the throng outside the duomo and walked towards the hills. As you go north up the main street towards the Valle dei Mulini the crowds gradually fall away and you can hear the surge of mountain water beneath your feet. This underground stream used to power paper mills that still line the street just before the town stops and the lemon gardens begin. Almost all the mills are derelict today and the stream rushes unhindered towards the sea. The street is also the site of a little restaurant called Il Mulino, where Giuseppina Orlando uses lemons grown by her brother Girolamo to make *calamarata* pasta and clams.

The dish is so pungent that the exquisite scent of lemons arrives at the table long before Girolamo can cross the room and deliver it. When I went to Il Mulino for the first time he was happy to recite the recipe as we ate. 'You'll need some clams, of course,' he said, 'and an Amalfi lemon.' He told me to cut the lemon into julienne strips and fry them with oil, garlic and chopped parsley. Add the clams, cover the pan and cook gently until they open. Now it's time to throw in the cooked pasta, but not just any old pasta. The ideal for this recipe is *calamarata*, a broad, soft, circular pasta, shaped like a giant's wedding ring.

I was so impressed by Giuseppina's *calamarata* that I try to visit whenever I can. I once arranged to drop in on my way back from the Aceto's farm with one of my groups of garden visitors. Giuseppina had devised a special, lemon-themed lunch for us, and the tables were piled high with radiant fruit. After the *calamarata* she gave us succulent sea bream wrapped in Amalfi lemon leaves and baked gently in the oven, followed by *delizia di limone*, an aptly named sponge baba, moist with limoncello and filled with a glorious chilled lemon cream. We all agreed that there could be no better way to spend a day in Amalfi.

One of the Sunniest
Places in Europe
Sicilian lemons, 'like the pale faces of lovers . . .'

﹍𑁋❦𑁋﹍

It's years since I approached Palermo from the water, and yet I still remember the beauty of its position, caught in the curved embrace of mountains and sea. Goethe's account of sailing from Naples to Palermo in 1787 was dominated by seasickness, but when the ship finally docked he was entranced by the vision of the city and the mountains behind it illuminated by afternoon sun. Instead of rushing down the gangplank to reach dry land, he lingered on board for so long that the crew eventually escorted him ashore.[1]

Despite the loveliness of its capital city, travellers' accounts of Sicily have always dwelt in equal measure on exceptional beauty and stinking urban slums; on glorious landscapes and the collections of 'dark and filthy hovels'[2] that counted as villages in the island's desolate interior; on the sophistication of the island's aristocracy and the base brutality of its bandits and *mafiosi*. Sicily has retained a reputation for organized crime, extreme poverty and general suffering, and yet anyone visiting Palermo for the first time cannot fail to be struck by the contradictory evidence of its magnificent buildings. Luxurious art deco villas cower in the shadow of modern high-rise blocks on Viale della Libertà; the ornate façades of churches and vast palaces tower over Via Maqueda and Via Vittorio Emmanuele, while Teatro Massimo, an assembly of imposing domes and columns built in 1876, was once the largest opera house in Europe. These handsome streets and buildings reflect periods of great prosperity, a prosperity due in part to citrus.

Everything about Palermo and its hinterland makes it well suited

SICILY

Palermo

Catania

N

Piana dei Colli •

Tyrrhenian Sea

CONCA D'ORO

Palermo ○

Monreale •

Ciaculli •

Bagheria •

Croce-verde •
Giardina

• Gibilrossa

to citrus cultivation. It is one of the sunniest places in Europe, and a great amphitheatre of mountains shelters it from the wind and interrupts the passage of clouds off the sea, obliging them to dump their warm rain on the land in winter. A half-hour drive from the centre of the city will take you to a swathe of land that has been known as the Conca d'Oro, or golden bowl, ever since the fifteenth century. The orange and lemon trees that grow in this fertile ground look like part of the natural vegetation, but they are really expatriates living thousands of miles from their places of origin. Lemons first grew as an under-storey in forests on the foothills of the Himalayas, and according to Tyôzaburô Tanaka, the great Japanese citrus expert of the twentieth century, all oranges came from Assam and Burma, where they were known as *naranga*, a name thought to derive from Tamil, in which the prefix *nar-* denotes fragrance.[3] However, more recent research by botanists working in the Chinese region of Yunnan has revealed so many primitive citrus forms that there's reason to believe many species of orange originated in China.[4]

Lemons and sour oranges reached Italy for the first time with the Arabs who invaded western Sicily in AD 831, coming ashore at Mazara del Vallo, on the island's south-west coast. Until then there had only been one kind of citrus in Europe, the citron brought to Calabria by the Jews in about AD 70. On Italian soil the name *naranj*, or *naranja* as the orange was called in southern Spain, evolved into *arancio* for the orange tree and *arancia* for its fruit. The word *arancio* is used as both a noun and an adjective in modern Italian, just as 'orange' is in English, but a noun can only function as an adjective if the object itself is familiar to the population at large, and it wasn't until the thirteenth century that Italians began to use *rancio* and then *arancio* as an adjective.[5]

At first sour oranges and lemons were grown only by the richest inhabitants, whose magnificent pleasure gardens were also a setting for trials of plants and trees recently introduced to the island. The aim of these experiments was to identify crop plants common to other parts of the caliphate that might be persuaded to grow in Sicily, and citrus trees were cultivated alongside other plants native to tropical and semi-tropical places, such as sugar cane, rice, aubergines, cotton and watermelons. The sour orange tree was one of the most robust of these new introductions, but lemons proved very fragile. They are hybrids, the progeny of cross-fertilization between their mother, the sour orange, and a paternal citron.[6] They can't stand frost, drought, excessive heat or rain and some people say their delicate nature is a throwback to their earliest existence in the wild, when they grew under the sheltering umbrella of bigger trees.

There was little about Sicily's natural climate to make the island suited to growing tender lemons, or even the hardier sour orange, as an agricultural crop. It delivered moisture only in winter, when, as Giuseppe Tomasi di Lampedusa remarked in his novel *The Leopard*, the downpours are '. . . always tempestuous and set dry torrents to frenzy, drown men and beasts on the very spot where two weeks before both had been dying of thirst'. And the parched summer, as Lampedusa also said, 'was as long and glum as a Russian winter'.[7]

If they were ever to thrive and be grown on a commercial scale, citrus trees and all those other exotic new plants needed the summer's heat and winter moisture to be delivered simultaneously, mimicking the hot, wet climates of their native lands. These conditions could be created by an efficient irrigation system, but when the first Muslim farmers arrived on Sicily they found only the crumbling and silted-up remains of the aqueducts, conduits and dams built between AD 211 and 440, when the island was a part of the Roman Empire.[8] This classical irrigation system could be repaired, but it was designed to capture winter rains and floodwater, which could be carried only as far as gravity would take them. Under Roman rule, Sicily's farming year had been defined by the availability of water, so it began in autumn, when the weather broke and the first rains fell, and finished in spring, when the water ran out. In summer the land lay fallow, parched and abandoned under the blazing sun.

The new irrigation system was a hybrid, a clever mixture of ideas from a variety of sources all over the vast landscape of the Islamic caliphate. What remained of the classical infrastructure was repaired, extended and integrated with a profusion of devices borrowed from sophisticated systems used to trap, channel, store, lift and distribute water in Egypt, the Yemen, Mesopotamia and the Levant. Every river, spring and lake was exploited, rainwater was captured in trenches on the sides of hills or mountains and *qanat*, long, interconnected tunnels, were cut through the bedrock below ground to harvest the water from aquifers.

Water was carried to wells or tanks through aqueducts, ditches, pipes or canals, and stored in artificial lakes, pools, reservoirs, dams and cisterns. It was raised to areas that couldn't be reached by gravity using a *noria*, or wheel, which was either hydraulically powered or turned by animals. The water was carried up in buckets or boxes attached to the circumference of the wheel. This combination of different methods and devices produced a more abundant and consistent supply of water than Sicily had ever known, delivering it throughout the year to a very much wider area of land than before, and producing crops with higher and more reliable yields.[9] The irrigated landscape reflected a high level of social organization between farmers, who had to make contracts or reach agreements about sharing the water available to them, about maintaining their section of the system and drawing on it only at an allotted time. Irrigation also altered the significance of agriculture on the island, making it a much more profitable occupation, and a more reliable source of employment.

~

The Arabs had brought both lemons and sour oranges along the Silk Road to Persia, where the orange's name changed to *naranj*. As the sour orange tree migrated across the Middle East towards Europe, recipes evolved to encompass the pungent and exciting flavour of its fruit. In Baghdad the sour orange became a key ingredient in *la naranjiya*, said to be one of the oldest Persian dishes still known to us today. Muhammad al-Khatib al-Baghdadi includes it in his thirteenth-century book of recipes.[10]

LA NARANJIYA

Al-Khatib begins by simmering coarsely diced lamb with chopped leeks, onions and carrots, adding cumin, coriander seeds, a cinnamon stick, ginger, pepper, ground mastic and a few sprigs of mint. When it's almost cooked, he minces the meat up with the seasonings and shapes it into medium-sized meatballs. And this is where the oranges step in, lifting the recipe into a new realm and transforming it into something distinctive and exotic. Having wiped the sides of the pan with dried mint, he strains the juice of three sour oranges into it. Obtaining the juice of a sour orange is a delicate operation in Al-Khatib's book. It must be done quickly by two people, one to squeeze and the other to strain the juice. Finally, he adds ground cardamom seeds to the orange juice before lowering the meatballs into it to simmer for a while before serving them with a sprinkling of dried mint.

Bal'harm, the city on the island's north-west coast that we now call Palermo, became the capital of Muslim Sicily. A first-hand account of the irrigated and perpetually productive landscape that surrounded it can be found in the *Book of Roads and Realms*, a description of a thirty-year journey through the Islamic caliphate made during the tenth century by a Muslim merchant and traveller from Baghdad called Ibn Hawqal. Palermo was his last port of call in 972. After that he returned to Baghdad and spent over a decade writing his book, finally completing it in 988. His account of Palermo is extraordinarily detailed. He was disgusted by most of what he saw of human life in a city where 'there are no longer any intelligent or able people, anyone with the slightest knowledge of science, or with any nobility of spirit or true religious feeling' – a situation he blamed partially on the inhabitants' mistaken habit of eating onions. Nevertheless, he was moved by the beauty of the landscape surrounding the city, the streams descending from the mountains to east and west, the water wheels lining their banks, and the land to either side of them planted with fruit trees, sugar cane, papyrus and pumpkins.[11]

On the outskirts of Palermo Ibn Hawqal found an abundance of gardens, all of them irrigated by water pumped up from the streams using *noric*.[12] It was here that the first orange and lemon trees had been planted. People from North Africa and the Near East used the word *paradisi* to describe these intimate garden spaces filled with the sound of running water and the perfume of beautiful flowering plants and trees, a shady sanctuary cut off from the harsh landscape outside by high walls. The beauty of the gardens was celebrated in a genre of poetry called the *rawdiya*, or 'garden poem', in which oranges and lemons were often mentioned. Abu al-Hasan Ali, an Islamic poet still living in Sicily under Norman rule at the end of the eleventh century, described oranges as pure gold that had rained on to the earth and been fashioned there into glowing spheres. Abd ar-Rahman, another Sicilian-Arabic poet, wrote:

> The oranges of the island are like blazing fire
> Among the emerald boughs
> And the lemons are like the pale faces of lovers
> Who have spent the night crying.[13]

Those wonderful Islamic gardens disappeared long ago, although Sicilian citrus groves commemorate their presence by being known still as *giardini* or even *paradisi* on the island's east coast, names that retain the echo of their Arabic associations with beauty, intimacy and succour, of the oasis in an arid desert landscape.

It's easy to spend all your time in Palermo on broad streets loud with traffic and lined with magnificent palaces, restaurants and enough baroque churches to keep you busy for a week. Venture off Via Maqueda and walk east towards the sea, however, and soon you will be in La Kalsa, a district that displays the marks of Arabic occupation in tightly packed narrow streets dating back to the ninth century. The name comes from *al Khalesa*, the Arabic word for 'pure', and whenever I've explored them, the streets have been quiet enough for gangs of small boys to play football and stray dogs to go about their urgent business undisturbed. There's often a smell of wood smoke, sometimes of refuse, and it's not hard to imagine yourself in the heart of Bal'harm, once the most intellectually advanced and civilized city in Europe.

Zagara, another Sicilian word derived from Arabic, has been adopted by the whole of Italy to name citrus blossom, and the specialist vocabulary of irrigation on the island has never shed its Arabic roots, so that a water tank is a *gebbia*, from the Arabic *gabiya*, a stream is a *favara* from *fawarra*, the mechanism used to pump water up from wells, rivers or tanks is a *senia* from *saniya*, and the raised channel that carries the water to the trees is a *saia* from *saquiya*.[14] This was the paraphernalia of citrus cultivation, and even now, when modern, pressurized systems and electric pumps are readily available, it's not unusual to find water managed in exactly the same way as it was over a thousand years ago.

The botanic garden in Palermo is an international centre for horticultural research and yet the citrus trees are still watered using the ancient Arabic system. And why wouldn't they be? It works. I once asked a gardener to explain the irrigation process to me. First he showed me the water being pumped up from underground. The pump was electric, but at ground level the water was released into a *senia*, a raised stone channel built to the traditional design. It rushed around the garden perimeter until it reached a small citrus grove. Then he opened a tiny sluice gate, releasing it into a temporary channel that he had raked into the dusty red earth at his feet. Caught in the sunlight, it gleamed like egg whites as it crept across the dusty ground towards the trees. When it arrived, he opened the side of the channel with his rake. Suddenly diverted, the water ran rapidly into a space enclosed by a small berm made from raked earth at the foot of the tree. The ground was parched and the water seemed to hesitate on its surface, as if it would rather run on, but as I watched, the ground gradually drank up the tiny lake and the lemon received its precious dose of liquid. The gardener stood for a moment, as if bemused by this close attention to the ordinary details of his day, and I paused too, delighted by the unbroken connection between the ninth and twenty-first centuries.

～

Irrigation was only one aspect of the agricultural revolution that made it possible to cultivate lemons and oranges. Their early evolution in the rich soils of the forest floor endowed citrus trees with a perpetual hunger and they would only thrive in fertile soils. However, much of the Sicilian landscape was steep, rocky and infertile and Moorish farmers had to make it more hospitable for their tender new arrivals by improving the structure and fertility of the soil.

The techniques they used are described in *Kitab al-Felahah*, the 'Book of Agriculture', a thirty-four-chapter treatise written by Ibn al-Awam in the latter part of the twelfth century.[15] The book combined classical knowledge with information from Arabic scientific literature and Ibn al-Awam's practical experience of farming his own land near Seville in southern Spain. He refers to more than fifty different fruit trees, including citrons, sour oranges and lemons, but whatever the plant or tree he describes, he always takes soil as his starting point. In the opening sentence he explains that 'the first step in the science of agriculture is the recognition of soils and of how to distinguish that which is of good quality and that which is of inferior quality. He who does not possess this knowledge lacks the first principle and deserves to be regarded as ignorant.' But his readers could not remain ignorant for long because he explained exactly how to distinguish between different qualities of soil by their texture, colour and the natural vegetation already growing on them. He was keenly aware of the importance of keeping the soil in good heart by maintaining its fertility and improving its structure. Soils with poor structure could be improved by mixing them with other soils. A heavy clay soil could be made more friable by digging in quantities of sand, and soil that was too sandy could be given more body by the addition of clay. He also advised mixing crushed bricks, tiles, rags and household compost into the soil to improve its texture. The addition of all manner of bird and animal dung, night soil, human urine or ground ivory would improve the soil's fertility.

The world was a perilous place for citrus farmers and Ibn al-Awam warned them to guard their trees against visits from menstruating women, as their mere presence would cause a total loss of both fruit and leaves. They could be protected from other intruders, such as camels or foraging livestock, with a liquor made from dogs' droppings or, 'even better because it will not be washed away', an emulsion of water and the fat of boiled puppies mixed with human urine.[16] It's not surprising that 'rags soaked in this mixture and tied to the trees' would 'repel animals through their rankness'. Sicilian trees were untroubled by foraging camels, but Ibn al-Awam's other techniques would have been well established in Sicily.

Muslim rule began to break down in 1060 when Count Roger de Hauteville, a Norman warlord and adventurer, invaded with an army of knights clad in heavy armour and mounted on powerful horses. Saracen soldiers, who were mounted on fine-boned Arab horses and wore light armour, were no match for them and by 1091 the Normans had conquered Sicily. They absorbed the best of Islamic science and culture and continued to farm and garden in much the same way as their Muslim predecessors, a situation that American author John McPhee described in his book *Oranges* in the following terms:

> The Norman conquest of Sicily turned into something of a scandal. Norman minds dissolved in the vapours of Muslim culture. Austere knights of Honfleur and Bayeux suddenly appeared in the streets of Palermo wearing flowing desert robes, and attracted to themselves harems of staggering diversity, while the Church raged. Norman pashas built their own alhambras. The Normans went Muslim with such remarkable style that even Muslim poets were soon praising the new Norman Xanadu.[17]

And it's true that Ibn Jubayr, an Arab born in Valencia and shipwrecked on the coast of Sicily in 1185, described the Norman kingdom and Palermo as 'an ancient and elegant city, magnificent and gracious, and seductive to look upon. Proudly set between its open spaces and plains filled with gardens, with broad roads and avenues, it dazzles the eyes with its perfection'.[18] Royal hunting parks full of deer, lakes, canals and gardens perfumed with fruit and flowers surrounded the city, 'just as a pearl necklace encloses a beautiful girl's throat'.[19] One of these lovely Norman gardens was made at Castello di Maredolce, 'Castle of the Freshwater Sea'. When it was first built in the tenth century, Maredolce, or La Favara as it was then, was an emir's *alcazar*. Water ran down from a mountain spring, forming a large freshwater lake that was remarkable so close to the sea. When the Norman king Roger II took the castle over, he renamed it and built a garden around the lake and several fast-flowing streams. Oranges grew on an island in the middle of the lake and Abd ar-Rahman, a Muslim poet in the court of King Roger, described them leaning out over the water, 'as if they wanted to see the fish that swam in it, and smile at them' and he compared their fruit to blazing fires among the green leaves of the trees. You may get a glimpse of Maredolce as you leave Palermo in the direction of Catania, although it's difficult to resist traffic that always seems intent on propelling you out of the city at high speed.[20] The empty castle looks vast, ancient and exotic among a sprawl of speculative buildings and industrial units that characterize the Mafia-controlled suburb of Brancaccio. The course of the stream that once fed the freshwater lake was diverted by the dual carriageway you will be speeding along, and there's nothing but rough grass and a few palm trees growing beside Maredolce today.

Norman rule ended in 1266 when Charles I of Anjou invaded the island and killed the king. Charles was an unpopular ruler and he was ejected in 1282 during the violent uprising remembered as the Sicilian Vespers. When his throne was given to Peter of Aragon, Sicily became an outpost of Spain and remained so for over 400 years. It became a backward and isolated place under Spanish rule, and the feudal system endured there until the British occupation of

1806–15. Much of the Conca d'Oro was divided into large estates, and during the sixteenth and seventeenth centuries the landowners spent the rents from their estates on building magnificent palaces to line the streets of Palermo and on buying aristocratic titles from the Spanish.[21] Nevertheless, in 1551 the Dominican monk and historian Leandro Alberti wrote about the Conca d'Oro as 'a fertile and agreeable landscape rich in charming gardens that are full of citrons, lemons, oranges and other delightful fruits planted in rows',[22] a description reminiscent of the Arabic gardens made on the same land 700 years earlier.

The citrus boom that endured in Sicily for over a hundred years can be traced back to the eighteenth century, and whenever I make the journey from Catania on the east coast to Palermo in the north-west, I am struck by the profound impact the cultivation of oranges and lemons has had on the natural landscape. Much of the journey is through the island's interior, where few trees have grown since the Romans first cleared the land and planted wheat. It is a vast, lunar landscape of semi-arid hills and anything that moves against that empty backdrop looks both tiny and theatrical. Glancing away from the motorway, I once watched a man trying to catch his horse, their two figures frozen in silhouette against an empty sky. Crossing the island on another occasion I saw only a shepherd on the flank of a hill, his sheep tied in a tight knot by vigilant dogs. But on the approach to Palermo, those bleached vacant spaces begin to fill with the lush colourful shapes of citrus trees. Many of them grow among high-rise blocks, or beside the broad arterial roads that surround the city, but the contrast between the island's empty interior and this cluttered fertile scene is as profound today as it was in the eighteenth century, when the market for sweet oranges and lemons began to grow steadily.

Fruit from the Conca d'Oro had always been sold on the mainland but now it was also exported to northern Europe. Sicilian lemons were used for a variety of purposes. Cooks in northern Europe had been using verjuice made from sour grapes, vinegar, quinces or crab apples to create the sour element in food, but as soon as lemons became available their juice supplanted all these options.[23] None of the tree's prolific crop was wasted. Even the poorest fruit could be squeezed for juice that would be exported as a concentrate and then processed to produce citric acid used as a flavouring and preservative. There was also a growing market in other lemon derivatives such as peel and essence used for flavouring food and drink, and essential oils extracted from the peel and used in the perfume industry.

By the mid-eighteenth century citrus production had become very lucrative for aristocratic families whose rights to land tenure on the Conca d'Oro dated back to the Middle Ages. The rumour of substantial profit attracted new investors to the area, some of them bourgeois families who had made good in the small towns of the island's interior and came to Palermo in search of a more sophisticated lifestyle and new investment opportunities. Others were merchants or lawyers who had made enough money to buy themselves aristocratic titles. They wanted to display their new status by investing in large estates.[24]

These new landowners poured their wealth into improving the soil and planting olives, citrus groves and vineyards, making the Conca d'Oro more beautiful than ever before. Olive trees and vineyards were established in the dryer parts of the landscape and citrus was introduced to areas with a reliable water supply. And now both old landowners and new built elegant villas for themselves against the dark backdrop of shining citrus trees or dusky olive groves in Piana dei Colli to the north-west of the city, Monreale to the south-west and on the ancient, feudal estate of Bagheria to the east.

To reach Bagheria today you take a road from Palermo that hugs the coast. I always enjoy the journey through the outskirts of the city, where the road passes a house surrounded by a moving haze of white doves. They belong to a man who hires them out for weddings. He releases them as the bride and groom emerge, and by the time he gets home they have already returned. These days the traffic crawls through ribbon development for most of the way to Bagheria, but as the road approaches the town the number of lovely eighteenth-century villas, magnificent gateways and long avenues scattered between the motorway, complex slip roads and citrus groves is quite extraordinary.

One of them is Villa Spedalotto, built towards the end of the eighteenth century and concealed inside the high walls of an enormous courtyard. The view from the broad terrace at the back of the villa is miraculously unbroken by new roads or buildings, so there is nothing to interrupt the great green sweep of citrus groves between it and the sea. An old pump house still stands on the edge of the groves, a memorial to the tedious lives of generations of donkeys that turned the wheel, or *noria*, inside it, pumping up water and sending it gushing into the two stone tanks outside. And in town, more villas and palaces made from pale tufa and warm limestone line the streets. They are light-hearted buildings, richly ornamented with carved balustrades, friezes and staircases. Some still have their own chapels and theatres, frescoed drawing rooms, mirrored ballrooms and beautiful gardens with terraces, fountains and long views over citrus or olive groves to the sea.

One building in Bagheria has always stood out from all the others. Villa Palagonia is behind a high wall in the centre of the town. When it was first built in 1715 it was normal enough, but that changed in 1749 when Prince Ferdinando Gravina inherited it. The first indication of the oddity of Palagonia comes as soon as you arrive in the garden, where rows of oranges grow alongside palm trees and cycads to either side of the drive, the colour of their fallen fruit startling against the red-ochre soil. Beyond them, the top of the circular garden wall is entirely filled by crumbling sandstone statues of dwarves and hunchbacks, soldiers, musicians, horses, dragons and an infinity of other strange figures. When Goethe visited in 1787 he described them as being '. . . designed without rhyme or reason, combined without discrimination or point, pedestals and monstrosities in one unending row . . .'[25] These days the magisterial stone staircase leads to a suite of empty rooms and terraces on the first floor, where the mirrors on the walls and ceiling of the ballroom have long since oxidized to black.

Goethe travelled all over Sicily, and like so many other English, French and German visitors on the grand tour, he was captivated by the beauty of the landscape. Many years later he encapsulated a universal longing among northern Europeans for the beauty, warmth and ease of life in the southern Mediterranean in a question that seems to haunt our collective imagination: 'Do you know the land where lemons grow . . . Do you know it well?'[26] Many of Goethe's contemporaries did know it well, and the Italian *vedutisti* who supplied grand tourists with paintings of landscapes, city views and studies of ancient monuments had already begun to include the Conca d'Oro among their subjects. *A Panorama of Palermo and the Conca d'Oro from Monreale* painted by Giovanni Battista Lusieri in 1782 shows the lush plain between the mountains and the sea dotted with walled citrus groves that are linked by straight roads to a scattering of barns and farmhouses, all of them built from the same golden stone. But this was a landscape in transition, and by the end of the eighteenth century the profits to be reaped from citrus had escalated into almost unimaginable wealth. Various events combined to push lemon cultivation on to an industrial scale, making it the first form of industrialized agriculture Sicily had ever known.

Antiscorbuticks

The beginning of the lemon boom came with acceptance by the Royal Navy in Britain that lemon juice, or *agro* as its bitter concentrate was called, could be used as both a preventive and a cure for scurvy. *Scorbuto*, as scurvy is in Italian, is caused by an acute deficiency of ascorbic acid, also known as Vitamin C. It is characterized by apathy, weakness and a series of repulsive symptoms including putrid and bleeding gums, loose teeth, foul breath, abnormal bruising and swollen limbs. Left untreated, the sufferer may die. Most mammals manufacture Vitamin C for themselves, but man, along with other primates and, surprisingly, guinea pigs, is obliged to consume it.

Anyone eating a normal mixed diet that includes fresh fruit and vegetables can absorb enough Vitamin C and so scurvy is most commonly associated with sailors on long sea voyages. They had known for centuries that it could be cured quickly and efficiently by eating citrus fruit, but it took nations – whose navies needed protection – a long time to agree.

Doctors observing the success of citrus in treating scurvy deduced it was the fruit's acidic content that exerted a therapeutic effect. Working on this basis, they began to invent complex alternatives to citrus fruit, and by 1696 the word *antiscorbuticks* (treatments for scurvy) had been included in the *Oxford English Dictionary* for the first time.[1] In 1747 a naval doctor from Edinburgh called James Lind undertook the first controlled therapeutic trial of lemon juice in the treatment of scurvy. Some people go so far as to describe the experiment as the first clinical trial in history. Lind was serving as ship's surgeon on HMS *Salisbury* in the Bay of Biscay. He recorded his own account of the experiment in *A Treatise of the Scurvy*, published in 1753:

On the 20th May, 1747, I took twelve patients in the scurvy on board the *Salisbury* at sea. Their cases were as similar as I could have them. They all in general had putrid gums, the spots and lassitude, with weakness of their knees . . . They lay together in one place, being a proper apartment for the sick in the fore-hold; and had one diet in common to all . . . Two of these were ordered each a quart of cyder a day. Two others took twenty five gutts of elixir vitriol three times a day . . . Two others took two spoonfuls of vinegar three times a day upon an empty stomach . . . Two of the worst patients, with the tendons in the ham rigid (a symptom none the rest had) were put under a course of sea water . . . Two others had each two oranges and one lemon given them every day. These they eat with greediness at different times upon an empty stomach. They continued but six days under this course, having consumed the quantity that could be spared . . . The two remaining patients took the bigness of a nutmeg three times a day of an electuary recommended by an hospital surgeon made of garlic, mustard seed, *rad. raphan.*, balsam of Peru and gum myrrh . . . The consequence was that the most sudden and visible good effects were perceived from the use of the oranges and lemons; one of those who had taken them being at the end of six days fit for duty . . . As I shall have occasion elsewhere to take notice of the effects of other medicines in this disease, I shall here only observe that the result of all my experiments was that oranges and lemons were the most effectual remedies for this distemper at sea.

Lind's trial should have put an end to scurvy in Britain's Royal Navy, but it took the Admiralty more than forty years to accept his findings and make it a legal requirement for sailors to protect themselves against the disease by taking an ounce of sugar combined with an ounce of lemon juice each day after two weeks at sea. The new rule had radical results. In 1780 alone 1,457 scurvy patients had been admitted to the navy's Royal Hospital Haslar, in Gosport, Hampshire, but in 1806 only two scurvy cases were recorded there.[2] At first lemons were supplied by Spain, but in 1798 Nelson conquered Malta and from 1803 onwards the Admiralty relied on Malta and Sicily to supply lemon juice to the British navy all over the world, making some people remark that Nelson had transformed Sicily into a vast lemon juice factory.

Lemon juice or lemons preserved in seawater were soon being used by the navies of several different nations to treat scurvy. However, in 1845 the Governor of Bermuda pointed out that the Royal Navy could substitute Sicilian and Maltese lemons with the limes grown on British plantations in the West Indies. This made good patriotic sense and by 1860 the Admiralty had abandoned their Sicilian suppliers and taken out a contract in the West Indies for the supply of Caribbean limes to the entire navy. By this time everyone agreed that lemons cured scurvy, but the British didn't differentiate between citrus species and 'lime' was often used as a generic word for citrus fruit of all kinds, so that the area where lemons were unloaded on the London docks was called Limehouse, just as it is today.

Nobody understood exactly why lemons cured scurvy and their antiscorbutic effect was still attributed to the acidity of their juice. Caribbean limes are highly acidic and so it was natural to assume that they would have a more powerful antiscorbutic effect than lemons. And yet citrus fruits are not interchangeable and fresh lime juice has only half the Vitamin C content of lemon juice. And to make matters even worse, the lime juice supplied to the navy wasn't even fresh: it was stored for long periods in open vats and passed through copper pipes. Unfortunately, Vitamin C is broken down by light, air and heat, and its destruction is catalysed by copper ions.

Thanks to the introduction of steamships in the mid-nineteenth century, the navy's fundamental mistakes had little impact on British sailors. The length of their voyages was reduced so dramatically that there wasn't time for them to contract scurvy before they were back on dry land. Nevertheless, the navy continued to administer lime juice, so that British sailors (and eventually all British people) were given the nickname of 'limeys'.

By now the only crews making long enough voyages to be at risk from scurvy were whalers and Polar expeditions. The crucial difference between lemons and limes was dramatically revealed in 1875, when George Nares led the British Arctic Expedition on an ill-fated attempt to reach the North Pole via Greenland.[3] The utter failure of this expedition was due almost entirely to scurvy, which afflicted both the sledge party on dry land and the men who had stayed aboard ship, obediently taking their dose of lime juice every day, as instructed by the Naval Medical Director General. This was a considerable embarrassment for the Royal Navy, which had boasted that it could keep a crew scurvy-free for two years.

It was a long time before anybody understood why lemon juice cured scurvy. When the role of vitamins in human nutrition was discovered, it was assumed that the active ingredient must be a vitamin, which was referred to as Vitamin C. Several groups of scientists all over the world attempted to isolate Vitamin C from lemon juice by carrying out repeated dietary experiments with guinea pigs. In 1927 a Hungarian biochemist called Albert Szent-Györgyi isolated a compound that he named hexuronic acid. A few years later he carried out his own experiment with two groups of laboratory guinea pigs. He fed one group on food that had been boiled to destroy its Vitamin C content and the other on food enriched with hexuronic acid. The group fed on the enriched food flourished, while the other guinea pigs developed scurvy symptoms and died. The experiment demonstrated that hexuronic acid was the antiscorbutic ingredient in lemon juice, and it was soon renamed ascorbic acid, though we generally refer to it as Vitamin C. Szent-Györgyi won the Nobel Prize for this work in 1937.

SICILY

Palermo

Catania

N

Piana dei Colli •

Tyrrhenian Sea

CONCA D'ORO

Palermo ○

Monreale •   Ciaculli •   Bagheria •

                          Croce-verde •
                          Giardina   • Gibilrossa

A Golden Bowl of Bitter Lemons

Extraordinary wealth on Sicily's west coast

⸻

The level ground between the city of Palermo, the mountains and the sea has been known as the Conca d'Oro for hundreds of years. Some people trace the name back to a poem in Latin by a Sicilian poet of the mid-fifteenth century called Angelo Callimaco. He described it as an *aurea concha*, a golden shell more lovely than the one Venus used to cross the sea.[1] But *conca* in Italian translates as 'bowl', and to my mind a golden bowl holds an even more potent promise of beauty and abundance than a golden shell. To the east of the city a steep winding road leads up to the village of Gibilrossa. Seen from this vantage point, the land looks like a generous green lap suspended between the two broad knees of the mountain, Monte Grifone. A coruscating blue sea hems it in on one side and on the other is the ragged sprawl of tower blocks that mark the edge of Palermo.

A century ago this sweep of land, from mountain to sea, was covered almost entirely by citrus groves. Today roads and modern buildings fragment the remnants of that green carpet, although from Gibilrossa you can still see a few of the high sandstone walls surrounding citrus groves on level or gently sloping ground, the citrus terraces cut into the lower slopes of the mountain, and a scattering of the lovely walled farmyards called *bagli*. It's a beautiful view, and yet during the nineteenth century some ugly and sinister events were played out in this gorgeous green theatre. The perpetrators of these crimes were a disparate group of men: bandits, smugglers and cattle thieves, and others with respectable roles as lawyers, politicians, estate managers and farmers. They were united by a desire for money and power, and they came together for the first time among the lemon groves on the Conca d'Oro to form an organization that would soon be known as the Mafia, a word derived from the adjective *mafioso*, meaning 'bold' or 'beautiful' in the dialect of Palermo.[2]

While Sicily's contract with the British navy lasted less than fifty years, the citrus boom continued unabated as an important citrus export business developed with the United States. The first shipment of lemons was sent to America in 1807 and by 1830 consignments of both oranges and lemons arrived in New York during every month of the year.[3] In 1832 America dropped the excise duty on Italian citrus and the quantity of lemons exported increased almost immediately from 3.5 million kilos to over 10.5 million kilos. By 1857 over 19 million kilos of fruit were making the journey across the sea. The most valuable fruit was carefully arranged in wooden 'American style' boxes and wrapped in brightly coloured tissue paper. The boxes were then slotted into gaps left among heavier and less palatable cargoes such as sulphur.

Sweet oranges grew alongside lemons, although growing oranges, which give only 400–600 fruits per tree, was never quite as lucrative as lemon farming. A lemon tree flowers several times a year and can be expected to produce between 600 and 1,000 lemons per annum, so that lemons generally accounted for two-thirds of Sicily's citrus production during the nineteenth century. The lemon tolerated the long Atlantic crossing to America better than the orange, although it was the orange that needed to be a good traveller, because it was generally eaten fresh and had to arrive in near-perfect condition if it was to fetch a good price on the American market.

A clipper could reach New York from Messina or Palermo in about forty-five days if the weather was good, but citrus is perishable, and if the ship was delayed by adverse weather, the whole cargo could be lost. The introduction of steamships in 1862 marked the beginning of a new and even more lucrative era, increasing the potential for profit and offering citrus merchants in both Palermo and New York reliable, prearranged departure and arrival dates for their cargo. And as the journey was so much shorter, the fruit arrived in better condition than ever before.

In 1860 Sicilian citrus production earned more money than any other agricultural activity in Europe. This was the year of unification, when Sicily shed its Bourbon rulers and joined the new Kingdom of Italy. It was also a time of political chaos in Palermo, and from this turmoil emerged the secret criminal organization now called Cosa Nostra or Mafia, a loose association of criminal groups that share a common structure and code of conduct. It's often assumed that this kind of organized crime was the ancient residue of feudal traditions that had evolved into something ugly among the most impoverished, isolated and backward inhabitants of Sicily. In reality many of the new *mafiosi* were aristocrats, and all of them were modern entrepreneurs who had become the most powerful landowners on the Conca d'Oro. The speculation, extortion, intimidation and protection rackets that characterize Mafia activity were first practised and perfected in the mid-nineteenth century among the citrus gardens of the Conca d'Oro, though they continue to blight politics, hobble the economy and cripple the lives of individuals on the island to this day.[4]

~

When Guy de Maupassant visited Palermo he saw it as a city '. . . surrounded by that forest of orange trees which has been called "the shell of gold" '. For him the trees were a 'black verdure [that] spreads like a dark stain at the foot of grayish and reddish mountains, which seem burned, wasted and gilded by the sun, so bare and yellow are they'.[5] The word *mafia* was already in common use by the time Maupassant was writing in 1885, although sometimes it was spelt with one 'f' and sometimes two, and there was general confusion about what it meant.[6] However, it was always associated with the Conca d'Oro, where large estates had traditionally passed from father to son. Now new laws had changed this ancient pattern of land ownership, creating fresh opportunities for investors, and soon the Conca d'Oro was divided among thousands of different owners.[7] Each of them had to spend enormous sums on buying plots of land, although the soil was often poor and stony. Gangs of labourers were needed to clear the stones, and sometimes the land had to be pre-planted with *opuntia*, the prickly pear cactus which has a powerful enough root system to break up rocky or compacted soil. Poor soil was improved by digging in huge quantities of topsoil and dung that had to be transported to the site over large distances. And before the new trees could be planted there were wells to dig and irrigation systems to install. Walls were needed to protect the young trees from thieves and cold winds, and each new citrus garden had to have an access road and a building where tools and harvested fruit could be stored. All this work represented a considerable investment, and it was a long time before you could expect to pay it off, because citrus trees don't bear fruit until they are three years old, it's eight years before they produce a proper crop, and many more before a new landowner could expect to make a profit.[8]

Such large investments and slow returns made landowners very nervous. They did their best to protect their trees, building three-metre-high walls around the land, training vicious dogs and appointing armed guards to patrol the walls at night.[9] And yet still they worried that somebody might cut off the water feeding the irrigation channels, or steal their crop before they could pick it. And if they did succeed in sending fruit to the docks for export, they wondered if it would be deliberately overlooked until after the ship had sailed. Such were the anxieties that gripped lemon growers on the Conca d'Oro and created perfect conditions for the first Mafia protection rackets.

The *mafiosi* were the richest landowners in the area. They offered to protect their neighbours' interests by using their own staff as wardens, workers and armed guards, and they were only too happy to supply water or install water pumps for farmers who could not afford to dig wells of their own. These *mafioso* water merchants also drew up obscure water contracts based on an impenetrable system of arcane traditions that allowed them to raise prices to extortionate levels whenever rainfall was low. It was also *mafiosi* who acted as wholesale fruit merchants and brokers, generally buying the fruit when it was still on the tree. (As soon as a sale was agreed, a single fruit was nailed to the outside of the door in the wall of the citrus garden and a shotgun cartridge was often attached alongside it as a warning to potential intruders.[10]) And finally it was the Mafia who controlled the hauliers who took the fruit to the docks and the dockers who loaded it on to the ships. By simultaneously creating risks and offering protection, the Mafia gained total control of every aspect of cultivation from the mid-nineteenth to the early twentieth centuries. And if anybody failed to pay their *pizzo*, or protection money, it was also *mafiosi* who dispatched gangs to scale the wall and hack down the trees during the night, or vandalize irrigation systems and cut off water supplies, intimidating staff and murdering anyone who resisted them.

There are several stories about brave people in this sinister landscape refusing to pay their *pizzo*; most of them end badly. Only one story has an unexpected and happy ending. In April 1867 a farmer called Ignazio d'Arpa refused to pay a *pizzo* to protect the water supply to a grove of Femminello lemons. At first the consequences of his refusal to pay were simple and predictable. The water was cut off at once and by July it looked as if the trees would die of drought. Cornered, the farmer paid up and as soon as the water supply was restored he drenched the ground between the trees and they responded to this sudden abundance by bursting into flower in August. Generally the Femminello can be harvested for eleven months of the year, with the only hiatus falling in August. This is bad timing because August is the hottest month, the holiday month, when sometimes only a lemon sorbet, a granita or a tall glass of iced lemon tea will do. And then, exactly a year after those drought-stricken trees had been watered and burst into flower, they were laden with the small green lemons that came to be known as *verdelli*.[11] These late-summer lemons fetched an exorbitant price, a price reflecting both their rarity value and the quality of their highly concentrated, extra-bitter juice, ideal for quenching August's thirst.

It has become standard practice to impose a deliberate drought on a proportion of Femminello trees in a lemon garden in April and to water them again only towards the end of July. By this means it has been possible to cajole and manipulate the Femminello into producing up to six crops of fruit each year. Each flowering and the harvest it produces have a different name. *Limoni invernali*, or 'winter lemons', are harvested between the end of September and the beginning of March. The most precious fruits in this crop are lemons that grow from the first flowers of the year (or *primofiore*) and mature at the end of September. April and May bring the *maiolini*. The lemons produced by the explosion of *zagara* in April mature between November and March. They are known as *bianchetti*. *Verdelli* come from a July blossoming that produces mature fruit in August the following year, and finally *marzani* and *bastardoni* mature in the autumn.

~

Citrus created an economic boom that began with the supply of lemons to the Royal Navy and endured until the beginning of the twentieth century, but the price of lemons or other citrus fruit could rise or fall dramatically from one week to the next, making the citrus markets more unstable than any other. Instability breeds risk and this encourages speculation. A *mafioso* citrus merchant would often buy the crop on the tree before it existed. Eventually he would arrange for a proportion of the fruit to be picked while it was still green. The green fruit kept well if it was stored in a cool dry place, and this allowed the merchant to hold fruit back, creating a shortage that drove up prices. When gummosis arrived on the Conca d'Oro in the 1860s, the Mafia turned even this fatal soil-borne disease to their advantage. Gummosis is caused by one or more species of the *Phytophthora* fungus, and it can affect the tree above or below soil level. Lemons grafted on to disease-resistant, sour orange rootstock are generally only affected above the graft, and the first symptom is sap oozing from small cracks in the affected bark. You might think the felling of thousands of diseased lemon trees a disaster, but the Mafia increased profits still further by selling lemons from the surviving trees at premium price.

Mafia monopoly over the citrus industry was maintained through violence and the beautiful landscape seemed strangely well suited to this purpose. The high walls of the lemon gardens served to conceal hired killers as they went about their work. Some of them stood on makeshift platforms behind the wall and shot over the top of it, while others made a hole in the stonework, just large enough for the barrel of a gun. When the job was done, they could escape quickly and easily by losing themselves in the labyrinth of lanes among the lemon gardens.

If you go to the Conca d'Oro today you will still find some of these narrow roads lined by walls so high you can see no more than the crowns of the trees that crowd up to them, occasionally flinging a branch over the top. I once arranged to meet a local farmer in a lane running between a derelict paint factory and the wall of a lemon garden. I stood, like a latter-day grand tourist, contemplating a lay-by full of junk, a small graveyard for white goods. Time passed and eventually I noticed that the narrow door in the wall of the lemon garden was ajar. The contrast between dazzling sunlight in the rubbish-strewn lane and the cool, ordered heart of the grove was acute, and I understood immediately why Sicilians continue to call their beautifully cultivated lemon groves *giardini*, or even *paradisi*. And yet this beautiful landscape once concealed something rotten, and if I had been foolhardy enough to take the same walk at the end of the nineteenth century I would have seen carved inscriptions and crucifixes on the walls to either side of me, marking the spots where people had been murdered for reporting a Mafia crime to the authorities.

In 1876 Leopoldo Franchetti and Sidney Sonnino, two intrepid academics from Tuscany, set out for Sicily to conduct an investigation into society and the criminal activity already known as *maffia*. Franchetti's part of the report, called *Condizioni politiche e amministrative della Sicilia* ('Political and Administrative Conditions in Sicily'), often draws attention to the contrast between the external beauty of the Conca d'Oro and the ugly activities common to a place where violence was 'exercised openly, calmly, regularly and as part of the normal course of events'.

In the lemon gardens Franchetti found 'every inch of ground irrigated, the soil hoed and hoed again, and each tree cherished as if it were a rare plant in a botanic garden', and admitted that anyone leaving Palermo for the first time to explore the countryside surrounding it would be enchanted. Indeed, if he were to forget every newspaper article he had ever read about Sicily's problems and go home at once, he might think it the easiest and most pleasant place in the world to live. 'But if he stays, if he opens a paper or eavesdrops on a conversation . . . everything around him will start to look different . . .' It's then that he'll hear about the warden of a lemon garden being shot from behind a wall because the garden's owner had appointed him for the job instead of the man the Mafia had chosen. And a little further down the road he'll see the place where a garden owner who decided to follow his own plans for renting out his lemon groves felt a bullet passing just above his head 'by way of a friendly warning'. And when you've heard a few stories like this, Franchetti said, 'the perfume of orange and lemon blossom begins to smell like corpses'.[12]

～

Until the First World War Italian lemons were used to produce the bulk of citric acid used in America. When war interrupted Italian exports, an American chemist called James Currie was driven to experiment with certain strains of a mould called *Aspergillus niger*. He found they would produce citric acid if they were fed on glucose or sucrose solutions, a discovery that would eventually destroy one of Italy's best markets for lemons. However, this disaster was masked by the fabulous success of a new introduction, the mandarin.

The sweet-fleshed, loose-skinned and highly perfumed mandarin is one of the three primitive fruits from which all cultivated citrus fruits are descended, and yet it was the last member of the citrus genus to reach Europe. The species to which it belongs is so large and complex that there has never been absolute agreement among taxonomists about the classification of all the varieties, hybrids and cultivars it encompasses, but clementines, satsumas, tangelos and tangerines are all derived from mandarins or their hybrids. Walter Tennyson Swingle (1871–1952), an American botanist and one of the most influential voices in twentieth-century citrus taxonomy, was what is known among taxonomists as a 'lumper', and consequently he united all the different varieties of mandarin under one species, *Citrus reticulata*, which he then divided into five separate groups. *Reticulata* derives from the Latin word *rete,* meaning 'net'. Ever since I learned this, I've found myself pausing between peeling and eating, just to appreciate the net-like structure of the pith that encloses the flesh of every variety of mandarin.

Swingle identified the Mediterranean mandarin that grows in Italy as a 'true mandarin', distinguishing it from clementines, satsumas and tangerines, which are also members of the mandarin group. But the Japanese taxonomist Tyôzaburô Tanaka, whose authority was equal to that of Swingle in the world of modern citrus research, took an entirely different approach. He was a 'splitter' and he subdivided mandarins into thirty-six different species.[13] In his system the Mediterranean mandarin was given its own name, *Citrus deliciosa*. Subsequent scientists found Swingle's system arbitrary and Tanaka's too complex, and indeed there is disagreement to this day. However, Italian citrus scientists, such as the late Francesco Calabrese, have always referred to the Mediterranean mandarin as *Citrus deliciosa*, just as Tanaka did.[14]

Other species had taken thousands of years to complete the laborious migration from their homelands in India, China and Malaysia, but the mandarin made an easy and efficient journey to Europe by boat. Sir Abraham Hume, a connoisseur with a passion for Italian Renaissance art, precious stones and natural history, had two seedlings sent from Canton in China to England by sea. When they arrived in 1805 he gave them to Sir Joseph Banks at the Royal Botanic Gardens in Kew. It was Banks who enabled the mandarin's migration to the Mediterranean by giving its seeds to a garden in

Malta, where they germinated and grew robust in the subtropical climate.

Some people say that Ferdinand IV, the Bourbon king, had the first mandarin brought from Malta to Sicily in 1810, and that he nurtured it at La Favorita, his magnificent hunting park at the foot of Monte Pellegrino. Other sources suggest that Giovanni Gussone, the director of Francesco I's Royal Botanic Garden at Boccadifalco on the outskirts of Palermo, imported ten trees from Malta in 1817.[15] Whatever the truth may be, by 1821 mandarins were already growing in the botanic garden of Palermo.

Citrus deliciosa was an immediate success in Sicily. By 1822 its sweet fruit had begun to be served as an exotic treat at the tables of Palermo's aristocracy. The new trees were a variety called 'Avana' (Havana), presumably named for the tawny colour of their pale golden fruit. 'Avana' produces a heavy crop every other year. Its fruit is quick to ripen and rich in essential oils, although it is full of seeds and deteriorates quickly if left unpicked. It took well to the mild climate and rich soils of the Conca d'Oro and in little more than a decade its fruit was produced in such abundance that it could be found in piles beside roads and bought for next to nothing.

In 1843 Sicily experienced its first true mandarin glut. The local market was saturated and the fruit could not be exported because its thin skin made it too fragile to endure a long sea passage – or even a short journey to the mainland and a bumpy ride by road. A tiny proportion of the crop could be preserved in brine or pickled in vinegar and exported to northern Europe, but most of it stayed in Sicily, where it was left on the trees to rot. Farmers responded to the problem of overproduction by felling their mandarins and regrafting lemons on to the original sour orange rootstock. Within twenty years many of those newly grafted lemon trees were fatally ill with gummosis, and farmers had no choice but to fell the diseased lemon trees and graft mandarins back on to the old sour orange rootstock.

It is this extraordinary versatility that has always made the citrus family so successful, and during the Second World War something almost miraculous happened in mandarin groves surrounding a little village called Croceverde Giardina, in the heart of the Conca d'Oro. I was told all about it by Salvino Bonaccorso, whose great-grandfather was at the centre of a story that begins during the mandarin harvest of 1939. When he harvested his crop at the end of November, he and three other local families noticed that some of the mandarins on their trees were still unripe. They left the unripe fruit where it was and when they returned to the fields in March 1940 to prune the trees they found it perfectly ripe. What's more, this unexpected, late-maturing fruit had far fewer seeds than the traditional 'Avana' variety of mandarin; its flesh was sweeter and more perfumed, and its skin was thinner and yet more resilient. The four families picked their mandarins and took them to market in Palermo. Fresh mandarins had not been available for two or three months and the fruit fetched such an exorbitant price that the events of that day are still remembered by the families' descendants. Salvino passed on his uncle's account of accompanying Salvino's great-grandfather to market with a cartful of fruit and returning with the same cart full of money. Suddenly those aberrant trees were so valuable that Salvino's great-grandfather took to patrolling the citrus groves with a gun, just in case anyone was foolish enough to try stealing the grafting material.

They needn't have bothered to protect their trees, because the same spontaneous change began to occur all over the mandarin groves of Ciaculli and Croceverde. This sudden transformation was due to a natural genetic instability, something that is common in cultivated citrus due to their hybrid origin and to propagation by cuttings, which can result in their shoots containing a mixture of genetically distinct cells. At a time when diet was still strictly con-trolled by the seasons, the significance of the trees' late-ripening fruit is almost impossible to overestimate.

Before long, all of the original 'Avana' mandarin trees were being replaced with the new late-ripening variety, which was given the very literal title of *mandarino tardivo di Ciaculli*, 'the late-ripening mandarin from Ciaculli'. The exorbitant prices fetched by their fruit persuaded farmers to extend the traditional boundaries of the citrus groves surrounding their villages. They used dynamite to break up the rocks before excavating new terraces from the steep sides of Monte Grifone, filling each one with topsoil and shoring it up with a dry-stone wall. This was back-breaking, expensive work, but they were spurred on by the knowledge the fruit would sell for incredible prices, to be eaten fresh or used to make ice creams, sorbets and marmalade.

~

The main street of Croceverde is still flanked by handsome houses built on profits from the citrus groves that still surround it, although these days it is a quiet place. Salvino Bonaccorso can remember his childhood, when pickers from all over the area arrived in the piazza of Croceverde on their bikes. 'The farmers would always test their biceps before hiring them,' he told me. His father would hire a gang of ten men to harvest the fruit and carry the boxes, and twenty women to cut off the fruit's sharp stalks. They all sang as they worked and sometimes there were singing competitions between Sicilian pickers and gangs hired in Naples especially for the harvest.

There were enormous profits to be made by mandarin growers when a day's labour cost only 700 lire, a kilo of *mandarini tardivi* fetched 1,200 lire, and it took just ten minutes for a picker to gather fruit worth the same amount as normal mandarins picked over the course of an entire day. Of course, these extraordinary profits were being reaped by the Mafia, who controlled every aspect of mandarin production. With the Mafia in charge, irrigation soon accounted for 50 per cent of production costs. Transport was another big expense, and of course the Mafia controlled that as well, making use of contacts that stretched well beyond Sicily and Italy to North Africa and America. And why was the *mandarino tardivo* such a lucrative crop? Because they fixed the price at an exorbitant level and kept it there, of course.

The two villages at the heart of the *mandarino tardivo* region were already traditional Mafia strongholds ruled by two branches of the Greco clan. The Second World War was over, but in 1946 a savage and bloody conflict broke out between the Greco in Ciaculli and those in Croceverde Giardina. Finally, the patriarchs of the Ciaculli branch, two elderly brothers, were murdered and the Croceverde branch emerged victorious. With peace restored in 1947, honour obliged the boss of the Greco from Croceverde to take responsibility for the orphaned sons of the Greco brothers murdered in Ciaculli. He gave them jobs in the mandarin gardens he managed in Croceverde and soon they were co-owners of a citrus export company, reaping exceptional profits from Ciaculli mandarins.[16]

～

The Second World War had left the centre of Palermo badly damaged by Allied bombing, and at the end of the 1950s the Mafia began to invest the profits from the mandarin groves and their other activities in a building boom so brutally destructive that it is generally known as *Il Scempio*, or 'the Sack of Palermo'. The Conca d'Oro, which had encircled Palermo for centuries like a beautiful green sash, was bulldozed to make way for uncontrolled ribbon development of jerry-built blocks that still overshadow a few remaining villas today, towering over them like a grotesque and gigantic parody of citrus trees. While numerous palaces and art nouveau buildings were demolished, the war-torn city centre was left untouched, and families were paid large sums to move out of their bomb-damaged properties to the new suburbs that now defaced one of the most beautiful cultivated landscapes in the world.

By the end of the 1960s cheap fruit had begun to arrive in Italy from South Africa, Spain and Israel, where production costs were reduced by mechanization, and from Tunisia and Morocco, where labour was less expensive. Production costs remained high in Croceverde and Ciaculli, where it was impossible to use modern machinery in the terraced citrus groves. And to make matters worse, the *mandarino tardivo* could no longer be sold at a premium because new varieties of mandarin that were seedless and easy peeling as well as late ripening had stolen its market. At this point the Mafia might have realized a long-term plan to build on their land in

Ciaculli and Croceverde, but their investment priorities had changed. Now the Greco clan found a new use for their mandarin gardens as cover for importing and refining heroin, a trade that would transform them from affluent businessmen into millionaires.

Heroin sold on the streets of America in the 1960s was imported from Indochina and refined in Corsican-run laboratories in Marseilles. When US President Richard Nixon declared war on drugs in 1969, these refineries were closed. Soon afterwards morphine base began to be processed in small refineries all over western Sicily, and it wasn't long before Sicilian *mafiosi* had almost complete control of refining, shipping and distributing heroin to the north-eastern United States, making the Sicilian Mafia richer and more powerful than it had ever been before.[17]

One of the most important refineries was hidden deep among the mandarin trees on La Favarella, the Greco family's estate between Ciaculli and Croceverde, where it's said there was a network of escape tunnels underneath the mandarin groves. La Favarella now belonged to Michele Greco, head of Cosa Nostra's governing commission, or Cupola, who was known as *il Papa* (the Pope) and regularly entertained bankers, cardinals, chiefs of police, politicians, aristocrats and businessmen at the farm. Giuseppe Barbera, now Professor of Arboriculture at the University of Palermo, told me about his own experience of *il Papa*. As a student he was given a grant to study the effects of a new irrigation system installed by Greco in the mandarin groves. 'I knew Don Michele well,' he recalls, 'and nobody worried about his Mafia connections in those days.' Apparently he dressed like a farmer in worn tweeds and liked to chat about football. He loved his mandarin groves passionately and always maintained that he was nothing but a simple farmer and a victim of malicious slander. Nevertheless, as Barbera is quick to explain, this benign figure was also the perpetrator of countless acts of chilling cruelty and repulsive ferocity. Such violent things happened in the narrow streets and piazzas of his territory, you might think blood oranges a better crop than mandarins in the fields that surrounded them.

Greco gave Barbera an enormous key to the main gate of the villa and a smaller one to the gate of the mandarin groves, so that he could go in and record data at any time of the day or night. However, Barbera often turned up to work to find that the lock on the gate to the groves had been changed and he was sometimes unable to get in for weeks on end. When Greco was finally brought to justice during the Maxi trial of 1986, his defence called on Barbera to tell the court about his research at La Favarella. They argued that a farm allowing free access to a student at any time of the day or night could not be the site of a heroin refinery, or any other illegal activity. Barbera told them about the changing lock, 'and that,' he says with great pleasure, 'was my contribution to *antimafia*'.

It's many years since Michele Greco died, but Barbera still has a few citrus fruit wrappers from La Favarella. He gave a couple to me and they are beautiful, unsullied things, emblazoned with the family name and an enormous butterfly on a gold background.

~

Conca d'oro is still used as a metaphor for fertility and abundance, and yet the real Conca d'Oro has become a strange liminal landscape, dissected by roads, disfigured by industrial units and tower blocks, and scattered with wrecked cars, shattered wardrobes and old fridges that have washed up like driftwood on the edges of roads and fields. In this context the mandarin groves that still surround Ciaculli and Croceverde have taken on huge cultural and historical significance. Some people, such as Giuseppe Barbera and Salvino Bonaccorso, have long recognized their importance as the last surviving fragments of the green belt that once acted as a lung, purifying and cooling the city's air.

When Leoluca Orlando was elected as anti-Mafia mayor of Palermo in 1993, he joined forces with Barbera, Bonaccorso and other local people to create a new future for the mandarin groves. In the past Ciaculli and Croceverde had been an impenetrable Mafia stronghold. However, Michele Greco had been in prison since 1986, serving multiple life sentences for dozens of murders, including those of policemen, politicians, civil servants and anti-Mafia magistrates. This was a political opportunity for Orlando, a means of penetrating the Greco clan's territory and transforming it. A magnificent and ambitious plan was devised to protect the groves and their fruit, and preserve the traditions surrounding its production while also bringing new employment to an area in desperate need. All this was to be achieved by transforming the mandarin groves into an agricultural park, open to all. They named their dream *Project Life* and the council began to draw up individual contracts with thousands of landowners. They offered to repay them for making their land accessible to the public by providing water for irrigation at a third of cost price and by promoting *mandarini tardivi* in the global market.

Between 1994 and 1997 a path was built through the mandarin groves and 6,000 new trees and shrubs were planted. There were proposals to convert abandoned farm buildings into accommodation and restaurants for tourists, and Barbera planned a botanic garden containing all the plants traditionally grown on the Conca d'Oro. Mandarin groves abandoned since they had been confiscated from the Mafia by the courts were brought back into cultivation. The ground on which these trees grew was thought to be cursed by the inhabitants of Ciaculli and Croceverde Giardina because it had belonged to members of the clans defeated in the terrifying conflict in the early 1980s known as the Second Mafia Wars. Other local farmers wouldn't touch it, but that didn't bother Salvino Bonaccorso and his colleagues, and they fetched their tools and got to work. Then the *mafiosi* in the community tried to undermine the project by suggesting that the park would be a kind of nature reserve where no one would be allowed to cultivate the land properly, or even prune their own trees, but the support of the mayor and the city council meant that even this cynical campaign could be overcome.

Project Life claimed years of hard physical and bureaucratic labour from its supporters, but it disintegrated as soon as Leoluca Orlando resigned as mayor in 2000. Salvino prefaced this final chapter of the story with a quote from that great Sicilian novelist Leonardo Sciascia. It wasn't a quote from one of his books. 'I saw him sitting on the city council one day,' Salvino told me, 'and suddenly he got to his feet and announced, "This city is irredeemable."' Today, Croceverde and Ciaculli are surrounded by signs of failure. There used to be a kilometre of open country between the two villages but now they are conjoined by dreary ribbon development. Many of the mandarin gardens are abandoned, the trees swamped by brambles and long grass, and in other places *case abusive*, houses built without planning permission, have sprung up among the trees. Between 2002 and 2007 an average of nearly forty-five hectares (over 110 acres) of land was lost each year to development.[18] 'And really,' Salvino concluded, 'there ought to be a sign at the end of the motorway saying "Abandon hope all ye who enter here."'

While mandarins, or 'easy-peelers' as they are known in the trade, are still bestsellers in the citrus industry, the *mandarino tardivo di Ciaculli* has become such a rare breed that it is protected by Slow Food, a campaigning organization dedicated to protecting and promoting local foods and small producers. My most vivid experience of the *tardivo*'s flavour was in a sorbet eaten at La Via del Sale, a Slow Food restaurant hundreds of miles north of Ciaculli in Turin. It was unforgettable, the flavour so intense it could be consumed only in tiny spoonfuls, each one leaving my mouth fizzing with the wonderful and unmistakable flavour of mandarin.

In the course of his lifetime, Salvino has seen the mandarin industry of Ciaculli and Croceverde change almost beyond recognition. His grandfather left ten hectares of citrus groves to his children. Only about two hectares are still cultivated today and that, according to Salvino, 'is a microcosm of what has happened to citrus cultivation all over Italy'. To reach his land we drove up a rough track, getting out just below the watermark of green trees on the rocky flank of Monte Grifone. Silence, heat, a scent of wild fennel and a view across the great bowl of the Conca d'Oro to the blank blue sea beyond. A few loquats grew among the citrus trees and we filled a couple of baskets with them before settling down in the compact shade of the mandarins. The trees had smooth bark and tiny pointed leaves. They were all tightly pruned into a goblet shape formed from three or four branches, so that each one seemed to embrace a space within it that was filled only by sunlight and the breeze. Salvino won't eat any other variety of mandarin. He lived in Paris for years, he told me, 'and once when I went back for a party, my friends bought me Spanish mandarins as a present'. Ill judged. 'I don't want to offend you,' he told them when he'd eaten one, 'but even a glass of water has more flavour than this.'

Later that day I had supper with Giuseppe Barbera and his wife, Margherita Bianca. They live on the north-west edge of Palermo, where mountains crowd the city, as if they were the invaders, instead of the high-rise blocks that have turned this area from open country to suburb in a generation. The house originally stood in a large garden among citrus groves that still marked the city's edge. Then an enormous road was built over the garden in front and tower blocks took the place of the orange and lemon trees, dwarfing even the *Ficus magnolioides*, a gigantic banyan tree that dominates the back garden. We sat outside as dusk fell and lights came on in tower blocks behind the banyan, and as we ate I asked Barbera for his prognosis on the future of Sicily's citrus trees. 'They will go back to being ornamental garden plants,' he said, 'just as they were when the Arabs first brought them here.'

A Sicilian Marmalade Kitchen

Marmalade attracts bigots. They believe in one true product made from the sour oranges the British call Sevilles, and coming most probably from a steamy Scottish kitchen in Dundee. But marmalade doesn't have to be made from oranges. The word derives from *marmelo*, the Portuguese for 'quince', and when it first appeared in fifteenth-century Britain it was a thick paste rather like Spanish *membrillo*, made from quinces flavoured with rose water, ambergris or musk and eaten at the end of a meal to aid digestion. The first marmalades came to England from Portugal, and later marmalade was also imported from Spain, North Africa and Italy. It was packed in circular wooden boxes and often imprinted with a design taken from the mould. Indeed, most European countries use 'marmalade', or *marmellata* in Italian, as a generic word for jams of all kinds, adding the name of the fruit to distinguish it from jams of other kinds. In Italy, jam made from oranges is *marmellata di arancia*. It is only in Britain that orange marmalade has a special status.

The earliest surviving British recipe for a marmalade made from Sevilles is 'a marmelet of oranges' in the recipe book of Eliza Cholmondeley, dated around 1677 and held by the Cheshire Archives. However, marmalade was not made on a commercial scale until the beginning of the eighteenth century, when stormy weather forced a Spanish ship laden with Sevilles to take shelter in the harbour at Dundee. James Keiller, a local grocer, bought the cargo at a very low price, only to discover that the oranges were sour, not sweet, and he was unable to sell them. His mother, Janet, had the ingenious idea of substituting oranges for the quinces she usually used to make marmalade. She sold it in the shop and it proved so popular that she began to make it every year. By 1797 the demand for orange marmalade in Scotland was so great that another Mrs Keiller and

another son called James opened the world's first marmalade factory in Dundee.

In Britain we think we have the monopoly on proper marmalade. I'd like to teach our marmalade fundamentalists a lesson by taking them to San Giuliano, an organic citrus estate on the eastern side of Sicily, outside Villasmundo and south of Catania, where they have been making excellent marmalade for years. We generally consider January to be the marmalade season in Britain, for this is when Seville oranges are imported from Spain, and yet at San Giuliano marmalade is made from November until May. Pot-grown citrus in central and northern Italy spend these months in the shelter of the *limonaia*, while the trees on the San Giuliano estate bask in winter sunshine, their roots deep in warm volcanic soil. Their fruit is so heavy it pulls the branches down, forcing them into the soft grass that grows with perpetual springtime vigour all over the orange groves. The air is full of the energy of ripening fruit, fruit that has swollen slowly, absorbing the heat of a long Sicilian summer. The orange groves look romantic, but San Giuliano is a working farm, an ancient *masseria* or fortified farmhouse, and although the snowy peak of Mount Etna can be seen in the distance, and the grass between the trees is full of wild flowers, beneath its lush surface the ground is rutted by tractor tyres, and if that earnest marmalade committee were to walk among the trees, they'd be as likely to turn their ankles in a rut as trip over ugly aluminium ladders and piles of brightly coloured plastic boxes, the practicalities of tomorrow's harvest. And wherever they went, they would be followed by a pack of friendly dogs, stray dogs that have found their way to the farm and the promise of regular food. They might notice the old Rottweiler, who would be a fierce-looking dog if he didn't make a habit of carrying a grapefruit with him everywhere he went. 'Grapefruit?' they might think. 'I hope they aren't silly enough to make marmalade from anything but Seville oranges. Heresy!' And yet that's exactly what they do at San Giuliano.

They began to make serious quantities of marmalade on the farm when Marchese Giuseppe Paternò Castello di San Giuliano and his wife, the late Fiamma Ferragamo, took over the estate from the marchese's father. Fiamma Ferragamo was world famous as partner and principal designer of the Ferragamo shoe label that she inherited from her father in Florence, but at San Giuliano a different side of her nature emerged. Leafing through family cookbooks in the 1980s, she was intrigued to discover a series of recipes for marmalade made from different species of citrus fruit grown on the farm ever since the nineteenth century. She soon began to experiment with the recipes in a rudimentary kitchen installed in a shed in the garden. She adjusted the original recipes in order to make marmalade in large quantities and, given her business contacts, it wasn't long before she began to export excellent, single-fruit marmalades to America and Japan.

Sadly, Fiamma Ferragamo died prematurely in 1998, but before her death she suggested that her daughter Giulia should continue with the marmalade business. Giulia will say she was very unsure about the idea, but she thought she would 'give it a chance'. She proved to have a natural talent and soon decided to build her own company. Under her control, production has moved to a professional kitchen on the edge of the citrus groves and expanded to include a wider variety of marmalades, organic honey, fruit slices preserved in syrup and a range of citrus-flavoured biscuits. Throughout the winter, perfectly ripe fruit is hand-picked each day and brought straight to the kitchen door. Inside, a small group of women make the marmalades entirely by hand. They might be working with lemons; tangelos (a mandarin–grapefruit hybrid); red grapefruit; clementines; mandarins, or sweet, sour or blood oranges.

Each marmalade is made to a slightly different recipe, the quantities of sugar and water being adjusted to suit the natural characteristics of the fruit. And over the years the marmalade makers have noticed that some fruits are easier to work with than others. On my last visit there, one of them remarked that you can't leave lemons alone for a moment because they will stick to the bottom of the pan. As she talked, she stirred continuously: 'They say I'd make a good baseball player,' she told me, pausing to show me the muscles in her stirring arm. But *tangeli* are a different matter. You can make marmalade from them almost without stirring and you can even get away with turning off the gas and heating it up again later. All of the fruit has to be washed, dried and cut up by hand. It is all organic and it is cooked without colouring, preservatives or added pectin. When it comes to deciding whether the marmalades are ready, the women do so by instinct. 'You can buy them all the thermometers you want,' Giulia says, 'but they'll never use them.'

When I watched the marmalade being made in the kitchen at San Giuliano I noticed that the overall amount of sugar was very much smaller than the quantities we are accustomed to using in Britain. Perhaps this is why each marmalade seems to encapsulate and intensify both the flavour and the vivid colour of the single fruit from which it's made. British marmalade snobs might refuse to be impressed by the news that San Giuliano marmalade is exported to America, but when they hear that it is now for sale in London's Sloane Square, perhaps even they might decide to try a jar.

Oranges Soaked in Sunsets

Blood oranges in the shadow of Mount Etna

⚬⚬⚬

The Conca d'Oro is not the only citrus-growing region in Sicily. There have been citrus groves for hundreds of years on the island's east coast, to either side of Catania and on the shores of the Strait of Messina. In Catania a glass of *selz al limone e sale* is still considered the ideal thirst quencher for a summer's day. You can order one at any of the beautiful art nouveau kiosks in the city centre and watch it being made while you wait. I once took a group of ten people to a kiosk on the edge of town. The young man worked like a machine, squeezing three half-lemons into a cup for each of us, topping it up with soda water and stirring in a teaspoonful of salt. It wasn't long before everyone realized that your first *selza* isn't necessarily a pleasant experience. Producers on the east side of the island enjoyed the same boom years as those on the west. Guy de Maupassant crossed Sicily in the spring of 1885 and remarked that the eastern shore 'exhale[d] such a powerful odour of blossoming orange trees that the whole channel is perfumed by it, as if it were a ladies' bower'.[1] He had reached a magic triangle of land on the plain beneath Mount Etna, a small area that produces the best blood oranges in the world.

~

One winter afternoon, years ago, I arrived in Catania just as the market was closing and the ground between the fish stalls was awash with icy water. Catania is a slower-moving, gentler place than Palermo. Most explain the difference between *palermitani* and *catanesi* by saying that people on the west of the island are descended from Arabs and those on the east from the Greeks. The Mafia operated the citrus industry on this side of Sicily just as it did on the other, but I'm told it has always had a different nature in Catania. If you were to open a new restaurant here, the Mafia would be at your door immediately. 'So you are opening a restaurant,' they'd say. 'What can we do to help?' You would know that all the bread you needed in a week cost €70 and yet, if you agreed to allow them to supply it for €100, that would settle the *pizzo*. They do business on the other side of the island in a different way: the Mafia would come to the door of your new restaurant in Palermo and demand a *pizzo* without offering anything in return. However, these distinctions have no relevance to the citrus trade today, because profit margins are so small that the Mafia lost interest long ago.

It was only two o'clock when I arrived in Catania on that winter afternoon, but the sky was so cloudy that lights were already coming on in the narrow streets. Caught in their glow I saw a pile of oranges. Some were split in half, their flesh the colour of blood, of garnets or old crushed velvet, and I recognized the fruit that Italian writer Carlo Emilio Gadda described as *arance imbibite di tramonti*, 'oranges soaked in sunsets'.[2] Looking back, I know the fruit must have been a variety called Moro, the bloodiest of the blood oranges produced on the volcanic plain surrounding Mount Etna, or, more specifically, on the triangle of land between Palagonia, Francoforte and Scordia, names that beg to be combined in a poem with Tarocco, Moro and Sanguinello, the varieties of *arancie rosse* growing there. Etna itself is often shrouded in mist, but when the weather is clear, its vast, snow-covered peak dominates every view and fills the windscreen of your car as you drive across the plain, and orange trees grow on an industrial scale all over the level ground.

The *arancia rossa* is a prince among oranges. The first written record of its presence in Italy comes from Giovanni Battista Ferrari in *Hesperides* (1646). He believed that a Genoese missionary had brought an orange to Sicily from China that tasted strangely like a grape and he remarked on its 'purple' flesh. This distinctive colouring is due to the blood-coloured pigments called anthocyanins that are also found in red, purple and blue 'super fruits' such as blueberries. The development of anthocyanin pigments in oranges is only triggered by a difference of at least ten degrees Celsius between day- and night-time temperatures while the fruit is ripening in the autumn and winter. In the shadow of Mount Etna it can be twenty degrees Celsius on a winter's day, but at night there is always a sharp drop in temperature. So it's cold, not warmth, that sets blood oranges on fire on the Etna plain. When they are grown in other places, such as Brazil or Florida, the contrast between daytime and night-time temperatures is unreliable, and coloration is often weak or altogether absent. This has made Sicily the most reliable source of blood oranges in the world.

Anthocyanins are good for us in a variety of different ways: experiments have shown that the anthocyanin content of blood oranges underpins high antioxidant activity, so that their juice, which is rich in Vitamin C, gives protection from certain kinds of cancer, increases insulin production, lowers the risk of heart disease and stroke, and improves circulation.[3] Plants use anthocyanins like a sunscreen, to protect themselves against ultraviolet light, and by eating plants or fruit rich in anthocyanins, we benefit in the same way.[4] An ongoing investigation into the benefits of blood orange juice is being carried out at CRA (Centro di Ricerca per l'Agrumi coltura), the Research Centre for Citrus Cultivation in Acireale near Catania.

I arranged to meet Dr Paolo Rapisarda of the CRA at a *mostra pomologica*, an exhibition of all the varieties of citrus grown in Sicily, both old and new. It was February and a violent storm broke as I set off down narrow lanes towards the farm. Water and bright red clay poured off the citrus groves to either side, filling the lane with such a quantity of fast-moving water that I wondered for a moment if I had taken a wrong turn and was trying to drive up a river. The event had been organized to remind local farmers of old citrus cultivars that had fallen out of favour and were in danger of extinction, and to introduce them to new varieties bred in Sicily, Basilicata and Calabria. The fruit was laid out in neatly labelled rows on a long table that ran the length of a marquee. Outside, the wind tore at the canvas and rain roared and hammered on the roof, but the dark space inside was filled with a warm citrus glow. Dr Rapisarda held his ground as the crowd of soaking citrus farmers pulled like a tide through the tent, and told me about his work at CRA-ACM.

Health-promoting organizations often recommend fruit juice because of its vitamin content, and orange juice accounts for almost 50 per cent of fruit juice consumption worldwide, but the habit of drinking fruit juice often goes hand in hand with obesity because of the high sugar content of sweet drinks. This problem has prompted Rapisarda and his colleagues at CRA-ACM to carry out further research on the effects of drinking orange juice. They used three groups of mice in an experiment designed to analyse the effect of juice from two varieties of sweet orange: the blood orange Moro and the Navellina, which would be described as a 'blond' orange in the citrus industry.[5]

All three groups were provided with a fat-rich diet and Group A was given water to drink, Group B the juice of blond oranges and Group C the juice of blood oranges. All the mice got terribly fat except for those in Group C. 'So what use is that, Dr Rapisarda?' I yelled, struggling to make myself heard above the roar of torrential rain on canvas. 'Ah,' he said, 'it means that by drinking plenty of blood orange juice we can eat whatever we like and stay thin.' Later he sent me the article he had co-written which explained that blood orange juice appeared both to limit weight gain and to confer resistance to obesity when compared to blond orange juice or water. In another experiment traffic police in Catania, whose job exposes them to high levels of pollution from exhaust fumes, were given ROC (Red Orange Complex), an extract obtained from Tarocco, Sanguinello and Moro blood oranges.[6] Asthma, chest infections and cardiovascular disease are all associated with air pollution and there is growing evidence to suggest that oxidative stress is one of the main factors in causing these conditions. After treatment with ROC, all but the smokers among the traffic police showed greatly increased resistance to these illnesses.

The high anthocyanin content of Sicilian blood oranges makes the fruit extraordinarily beneficial to those who drink it, and yet the trees can only be relied on to produce anthocyanin-rich fruit if they are cultivated in the perfect conditions of the tiny area of eastern Sicily surrounding Mount Etna. So very few people have access to blood oranges that Professor Cathie Martin and a team at the John Innes Centre in Norwich, England, have joined forces with the CRA-ACM in Sicily and the Sichuan Academy of Agricultural Sciences in China in an attempt to change this situation. They have engineered the Ruby gene responsible for the pigmentation of blood oranges so that it does not need contrasting day- and nighttime temperatures to develop.[7] If the experiment is successful, it is possible that the Ruby gene could be engineered to be active in blond fruit. This would mean that blood oranges could be produced in far greater quantities, and their health-giving properties made very much more widely available. Unfortunately, it would also break Sicily's monopoly of premium fruit in the blood orange market.

In Italy blood oranges have become something of a symbol for healthy eating. This is due in part to the invention of a brilliant scheme by Princess Borghese at Il Biviere, an organic citrus-growing estate near Lentini, about thirty minutes' drive south of Catania. Today the farm is in an area that offers the best conditions in the world for producing blood oranges, and yet when Miki and her husband, Prince Scipione Borghese, first arrived in 1968, it was a godforsaken spot with a long history of suffering. Until 1931 an enormous lake covered the entire area and Giovanni Verga, who was born in Catania, wrote vividly about the miserable lives of the peasants living on its shores in a story simply entitled 'Malaria', published in 1883:

> In vain the villages of Lentini and Francoforte and Paternò try to clamber up like strayed sheep on to the first hills that rise from the plain, and surround themselves with orange groves, and vineyards, and evergreen gardens and orchards; the malaria seizes the inhabitants in the depopulated streets, and nails them in front of the doors of their houses whose plaster is all falling with the sun, and there

they tremble with fever under their brown cloaks, with all the bed-blankets over their shoulders.[8]

In 1931 the lake became a focus for Mussolini's campaign to rid Italy of the wetlands that provided such a perfect habitat for malarial mosquitoes. It took twenty years to drain all but the small area of water that remains to this day, and when Prince Borghese inherited Il Biviere, the house was still surrounded by a bleak post-diluvian landscape. If you ask Miki what it was like when she moved there with her four small children, she will tell you the house was an awful place, derelict and rat-infested. 'And there was no water, no trees or bushes. In fact, there was absolutely nothing here at all.' And yet the soil, enriched with a mulch of water plants and stranded fish, was exceptionally rich, and today Il Biviere is surrounded by a wonderful garden made by Miki and her husband around the ancient Roman quays of the harbour, where mooring ropes have worn deep grooves and even holes in the stone. A small door leads from the shady harbour to the dazzling sunshine of the main garden, scattered with the water-worn lumps of porous rock that once lay on the lake floor. Over the years, the Borghese planted an extraordinary collection of palms, ornamental trees and succulents, creating mixed stands of exotic and native species, so that the garden sits comfortably in the surrounding landscape.

Twenty years ago the Borghese were the first citrus farmers to go organic in Sicily, something that's virtually standard practice today. And nearly twenty-five years ago Miki invented a scheme that has benefited organic blood orange farmers and AIRC (Associazione Italiana per la Ricerca sul Cancro), Italy's cancer research charity, in equal measure. The charity is assisted by government sponsorship to buy huge quantities of organic blood oranges every year. The oranges are packed in nets and dispatched to AIRC volunteers all over the mainland. On either the last weekend of January or the first of February, volunteers sell nets of blood oranges and give out information about healthy eating from stalls in the main piazzas of 2,000 cities. The scheme has gone from strength to strength and in

2012 they sold 400,000 nets of oranges, raising €3.7 million of clear profit for cancer research.

I often take groups of British garden visitors to see Miki and her garden, and that's where I am used to meeting her, hospitable and elegantly dressed, introducing us to each extraordinary palm tree, cactus or succulent plant as if we were all friends at the same party. In the packing house where Il Biviere's oranges are processed, Miki cuts a very different figure. She dresses in ski boots, a thick jacket and ear defenders, the right kit for this cold, illuminated hangar, where fork-lifts dash about, every move accompanied by their peculiar language of squeaks and grunts. The machinery for sorting, washing, drying and packing the fruit creates another layer in the cacophony and, at the centre of it all, a man with a microphone stands like a bingo caller on a raised gangway, yelling out the details of orders. All around him is citrus fruit in motion, fruit in a hurry, moving like a gold ribbon along conveyor belts and jiggling companionably together under the washer.

That's the end of preparation for Miki's organic produce, although non-organic fruit would then be polished in a stinking mixture of wax, fungicide and ammonia and dried in a roaring tunnel of hot air. Finally the fruit is dropped through gauges that sort it by size before it cascades at last into the stillness of a net or box, the most processed of all unprocessed foods.

～

At the height of a Sicilian citrus harvest in January, February and March, the average packing house sorts and packs 1,500 tons of oranges a day. Very few of these Sicilian oranges ever find their way to British supermarkets, where the fruit counters are dominated by bland and oversweet oranges from South Africa, Brazil, Morocco, Israel and Egypt. A common orange, like the Navel or Valencia, has a sugary, one-dimensional taste. Eating a Sicilian blood orange is a much more complex experience. Take the Tarocco: its meltingly soft flesh also has a high sugar content, but its sweetness is balanced by high acidity. The result is a complicated, multi-dimensional flavour that unfolds slowly, subtly, beguilingly, making any other kind of orange seem sharp, cloyingly sweet and intolerably crude.

Cooks in Sicily make the most of the Tarocco. Wafer-thin discs of its marbled flesh are combined with fennel, good olive oil, salt, a sprinkling of chopped fennel leaves and black pepper, or used in pale pink risotto, bright red jelly and dark pink ice cream. Tarocco peel is also candied, a process that takes three days to complete.

CANDIED TAROCCO PEEL

The first step is to cut the peel into fine strips. Now soak it for twenty-four hours in salty water. Next rinse and soak it again for forty-eight hours in fresh water that must be changed morning and evening. Finally, put the peel into a pan with its own weight in sugar and barely enough water to cover it, and cook over a low flame until it is almost dry. The candied peel must be put into a sealed jar while it is still tepid.

We are regularly denied these wonderful experiences because blood oranges are expensive in comparison to mass-produced oranges from other places, their season is short and their skins delicate. Supermarkets prefer to fill up on the cheap fruit they can make available to their customers throughout the year, and they particularly favour Valencias because their hard skins stand up well to supermarket treatment. If you want to find blood oranges outside Italy, your best bet is to seek them out in independent shops that sell organic produce.[9]

～

Miki can remember a time, barely forty years ago, when a single hectare of Sicilian orange trees would fetch the same astronomical price as a hectare of land among the Tuscan vineyards producing

Brunello di Montalcino, one of Italy's most valuable wines. The Italian government used to protect the citrus industry from imports, but in the mid-1990s the controls were abandoned and the Italian market was flooded with cheap blond fruit from abroad. As a consequence of this change in policy, over 30 per cent of Sicily's citrus groves have been grubbed up, and each year things get a little harder for citrus farmers.

Cristina di Martino, an Italo-American journalist whose Sicilian grandparents emigrated to the United States, is the only member of the family to have returned to Sicily, where she lives in Catania and writes about food and trade for professional magazines. Her Sicilian genes emerge in her cooking, and when I went to her apartment we ate royally at a table in a high room where she lit a charcoal brazier to warm the winter air. The story of rediscovering her grandfather's house and meeting her Sicilian cousins for the first time made her cry in the telling, and I could have cried too when she shared her views about Sicily's citrus industry. She described it as an industry at risk, vulnerable both to the myriad diseases afflicting citrus trees and to the ever-expanding cultivation of citrus in other countries. 'Watch out for China,' she said, and she wasn't the first to tell me that Sicilian growers should stop competing and start cooperating. Until then they will have no power to negotiate with supermarkets that habitually pay pitifully low prices for fruit they then retail at a mark-up of between 500 and 700 per cent. And to make matters worse for growers, the supermarkets don't pay their bills for five months. Big farms struggle with these conditions and inevitably many of the tiny scraps of land used for generations by one family to grow oranges have been abandoned.

~

Small orange groves were often farmed uneconomically for decades and their loss is of little financial significance to the island as a whole. Nevertheless, the very characteristics that made these farms old-fashioned and ultimately inefficient were of enormous cultural value. The fruit was often grown on marginal land and steep sites terraced hundreds of years ago. When cultivation becomes unprofitable, these difficult, labour-intensive plots are always the first to be abandoned. They deteriorate quickly and their decline represents the loss of an important feature of the agricultural landscape. Citrus production is more lucrative on the broad plain surrounding Mount Etna, where trees can be cultivated on an industrial scale, and yet the transition from small family holdings to industrialized cultivation represents a threat to the ancient skills and traditions that have accumulated around citrus over many centuries.

The citrus harvest starts in Sicily just as winter begins to make deep inroads into autumn and deciduous trees are hurrying towards dormancy. That's the time to go to the citrus groves, where sunlight drives the throbbing engine of ripening fruit and the harvest is about to begin. A few years ago I tried to get work in Sicily during the orange harvest. Every farm I asked simply said no, and then someone explained that the orange crop is sold to a citrus merchant while still on the tree, just as it was on the Conca d'Oro in the nineteenth century. The merchant or the packing house will send their own team of pickers and so the farmers had no power to offer me work. Eventually I was put in touch with Rudolf and Benedikta von Freyberg, a German couple who have lived for decades at San Giorgio, another organic citrus estate on the plain between Mount Etna and the sea. Just like Il Biviere, San Giorgio is on the fertile ground once covered by the lake, and the peak of the volcano is the backdrop to each long view across the von Freybergs' land. Benedikta admitted she'd no idea if the pickers at San Giorgio would let me join them. They were a team like any other, sent by the local packing house, but she said I was welcome to come when the harvest was at its peak in February.

The von Freybergs' house is surrounded by 120 hectares of citrus groves that lap like a green sea around the edges of its garden. These are not the walled or terraced citrus gardens of the Conca d'Oro; they are huge open expanses of level ground covered in neat lines of trees. I'd arrived in time for the Tarocco harvest. The Tarocco is a lovely compact tree that is usually planted in rows that run from north to south, ensuring that the sun falls evenly on its branches to ripen the fruit. The trees are regularly pruned, so they never grow very tall and light can penetrate to the centre of each tree, encouraging it to fruit as well on the inside of the branches as it does on the outside. The suckers, new shoots that come straight from the rootstock, also have to be removed during pruning. In Sicily suckers are called *bacchettoni*, and I'm told this is a word you might also use to describe your sister's boyfriend if he were tall and handsome but utterly useless in every other way.

Tarocco trees grow alongside Valencia oranges at San Giorgio. The Valencia is a huge, tough tree that shrugs off disease and can be relied on to produce heavy crops. Leave it unpruned for years and it will continue to fruit abundantly. And yet its fruit is a crude, nasty thing in comparison to the Tarocco, a great big ball of sugary juice fit only for squeezing. The Tarocco has an entirely different nature: it is delicate and demanding; ignore it and it will fall sick; neglect pruning for two years and it will produce a scant crop of inferior fruit.

It rained early on my first day at San Giorgio and I learned the truth of a popular saying used all over Italy, *Il buon contadino va in cantina quando piove*, which translates roughly as 'When it rains the sensible farmer goes to the pub', and so I wasn't surprised when all the orange pickers went home. However, when I woke at seven the following morning I could already hear a slow procession of lorries and minibuses advancing up the drive. I scrambled out of bed and into as many clothes as I could find – it was a windy, overcast morning, and it's true about the cold nights down there, in the shadow of the volcano. I soon realized that I needn't have hurried, because starting work in an orange grove is a slow process made up of well-defined phases. There was no question of piling out of the bus and getting on with the job. First the coffee had to be made, for who could work without it? When I arrived, the men were still standing in quiet groups, gazing at the *macchinetta di caffè* balanced precariously on a camping gas burner in the back of a minibus. The pickers were men, all of them Sicilian, because orange picking isn't a job given much to foreigners, although a few Romanians have been accepted into some teams.

We stood about in dawn light, sipping our coffee from tiny plastic cups. 'What are you up to?' they wanted to know. 'Why are you so interested in orange picking?' and, more importantly, 'Did you actually know Princess Diana?' As soon as the coffee was finished one of them set off purposefully towards the trees. 'Hey!' everyone shouted, for there were cigarettes to be smoked before picking could start. And then everything happened at once.

A lorry roared up and piles of brightly coloured plastic boxes were thrown out of it on to the ground. Everyone grabbed a bucket attached to a broad strap and slung it over one shoulder. They pulled on gloves and soon they had propped up their ladders and settled in among the branches of the trees. At first it looked as if the pickers had magic tips to their fingers. They seemed only to pass a hand across the fruit to detach it from the tree, but then I caught a glimpse of the tiny secateurs concealed in their palms. 'It's not child's play,' they said. 'New pickers usually cut themselves.' Someone sang, someone swore, everyone told jokes and they all worked at frenetic speed. The trees grew close together, it was an enclosed and private world beneath their branches, and soon the pickers had disappeared among them. Sometimes I'd glimpse a woolly hat emerging from the leaves at the top of a distant tree, then it was gone. There were fallen oranges on the ground. Some had already been mined by insects that left holes in their sides, revealing red flesh like gaping, gory wounds.

One or two trees at San Giorgio had succumbed to Citrus Tristeza Virus (CTV), also known as quick decline, which is one of the most destructive diseases known to the citrus industry. Over the centuries the sour orange, so often used as rootstock, has proved resistant to both gummosis and mal secco disease, but it has little or no resistance to this scourge. CTV originated in China and has spread across continents through the sale of infected bud wood (cuttings for use in grafting). It is spread locally by a variety of different kinds of aphids. There are several strains of the virus and the symptoms of CTV depend on which strain has infected the tree and the climatic conditions in which it grows.

The most significant symptom from an economic point of view is the rapid decline or death of trees grafted on to sour orange rootstock. Millions of trees have been affected in southern Italy and Sicily, and it's not uncommon to see entire groves of dead trees, regiments of grey ghosts haunting the fields. Many citrus groves have already been grubbed out and replanted with trees grafted on to alternative rootstocks, such as *Poncirus trifoliata*. *Poncirus* is a member of the Rutaceae family, like citrus, but it is frost hardy, and so far it has proved entirely resistant to CTV. It is becoming increasingly popular as a rootstock. I didn't stay very long with the orange pickers. Despite all my determination to work during a harvest, once there I felt I was only playing, while all around me men were working hard. They worked as a team to clear the fruit from a specified area of trees. The faster they worked, the sooner they could go home.

On my last night at San Giorgio I returned to the orange groves at dusk. It had been windy ever since I arrived on the farm and it was still windy, very windy. I took the track leading past some Valencia orange trees. Their fruit ripens late and, unpicked, it glowed among the dark leaves like the 'golden lamps in a green night' of Andrew Marvell's famous poem, and seemed to radiate light into the stormy winter dusk.[10] Crows settled noisily in a group of old stone pines. The only other sounds in the huge silence were of wind high in the trees and the continuous croaking of frogs. I took a stony track up the hill towards the last of the light. At once the wind freshened, blowing full in my face. A dog had followed me from the yard and he trotted at my side. There were caves in the hillside above where the Siculi, prehistoric inhabitants of eastern Sicily, used to live, and on the steep ground rocks lay exposed, as if the island's ancient bones were breaking out through the thin soil. That's the soil the Tarocco loves, the soil it gathers to itself, so that Sicilians believe a Tarocco orange is steeped in the flavour of this mysterious, ancient place. Even if the attempt to engineer the gene responsible for the pigmentation of blood oranges succeeds, it seems impossible that the new generation of fruit could have the same unforgettable taste.

Suddenly we emerged, the dog and I, into a grove of seedling oranges on the hilltop. A hare cantered lazily away from us and the dog tore off in a chaotic and hopeless pursuit. Behind me Mount Etna's snowy peak loomed, phantom pale. Ahead the open hills were windswept and empty against the twilit sky. I walked around the perimeter of the orange grove on a rough track, and soon it was almost too dark to see. As I began the descent towards the farm, the dog fell in with me again, panting after a long chase. By the time we reached level ground it was too dark to see him, but I heard him pause to lap water from a puddle. The orange lanterns were extinguished.

The Runt of the Litter
Liguria's cosseted chinotti

⁓◦◦◦⁓

These days citrus is produced on a commercial scale only in Sicily and southern Italy, but over the centuries it has made an impact on a very much wider area of the country. I stumbled across a metaphor for the broad geographic span of citrus in Italy on an evening in Genoa, capital of Liguria. It was winter and yet the climate on the coast is so mild that this region of north-west Italy was once as important as the south for citrus production. The prostitutes hovering in the narrow streets near the port were wearing shorts, and almost all the tables outside bars and restaurants were taken. Eventually I found a seat overlooking the old docks and ordered a Campari and orange. The waiter looked bemused and I felt the need to justify myself, explaining I'd often drunk Campari like that, and thought fresh orange juice and ice its ideal companions in a glass. His face lit up and he said, 'Ah, we call that a Garibaldi!'

I could see the barman mixing my drink. It was a slapdash affair, a question of two Sicilian blood orange halves squeezed messily into a tall glass, the glistening juice escaping through his fingers. He filled the remaining space with Campari, a fistful of ice and a twist of orange peel cut as long and fine as a shoelace. When the waiter returned with my scarlet drink it was suddenly obvious that its name came from the colour of the shirts worn by the *Camicie rosse*, soldiers in Garibaldi's army of a thousand volunteers.

While Garibaldi was fighting to make Sicily and its blood orange groves part of the new Kingdom of Italy, Gaspare Campari was in the northern Italian region of Lombardy developing a recipe for the aperitif that would take his name. Lombardy includes the western

shore of Lake Garda and, despite its latitude, it was once the centre of a thriving citrus industry, producing lemons that were exported all over northern Europe. Finally, an extract from an exceptionally acidic citrus fruit called *chinotto* is one of the ingredients used to give Campari its bitter taste, and the history of the *chinotto* is entwined with Liguria, the other great centre for citrus production in northern Italy. This made my drink both unification in a glass, combining north with south, and also a metaphor for the extraordinary importance that citrus cultivation once had across the whole of the Italian peninsula, from Lombardy in the far north to Sicily and the south.

Though citrus has been a lucrative business in small areas of both Liguria and Lombardy, today cultivation on a commercial scale is little more than a distant memory. Nevertheless, you will still find lemon trees growing on the west shore of Lake Garda and on every patch of level ground between the steep streets and squares of Liguria's Riviera towns. When work took me to Bordighera on the Riviera in July, lemon trees all over town were laden with pale fruit. One was weighed down by such a heavy crop that its owner begged me to pick all I could carry. It was late in a day of blazing heat and the lemons were heavy and hot, small suns radiant in my hand.

Lemons have always produced fruit in these industrial quantities on the Riviera, where a combination of the sea, which prevents the temperature sinking too low, and the Apennines, which shelter the coast from cold winds, creates a peculiarly mild microclimate. However, citrus has a dangerous existence here, for although the climate is mild enough for the trees to thrive outside all year, every so often there is a disaster such as the great frost of 1956 that destroyed the bulk of Riviera citrus trees. Lemons grown in these cooler climates were especially bitter, a characteristic that made them particularly popular in northern Europe. These lower temperatures were less helpful to oranges, which were always a little sour in comparison to Sicilian fruit.

~

There are very few citrus farms left in Liguria today and so I was lucky to meet the Parodi, a family who cultivate citrus on their smallholding near the seaside town of Finale Ligure and sell the fruit at local markets. I reached their land by driving through narrow streets late on a November morning, just as the doors opened at the end of Saturday school. I had to inch the car upstream against a powerful current of jostling, laughing, oblivious adolescents and inexpertly driven mopeds, before stopping at the head of the Valle dell'Aquila, a steep green valley running inland from the sea. The Parodi smallholding was on the other side of the river, a jumble of water-torn stones surrounding a benign trickle of water. Not for long, I thought, glancing up at the snowy mountains beyond. On its far side I found a narrow path bounded by walls that took several dog-leg turns, as if trying to shake off the citrus trees crowding against it, their dark crowns already studded with fruit.

Giacomo Parodi was waiting for me at the gate of the smallholding he inherited from his father thirty years ago. He was a large man with a dark beard, and when he greeted me, my hand disappeared entirely inside his. He had passed the farm on to his son, Alessandro, but that seemed to make no difference to him or his wife, who appeared to be working as hard as ever.

It didn't take long to realize that I was in a kind of paradise. Everything was golden or honey-coloured in that ancient and intensely cultivated landscape: the sheltering walls of the citrus grove, the beautiful villas that seemed to grow organically from the surrounding hills, the warm air, the autumn leaves of vines that striped the sides of the valley and the bright fruit on the trees. The trees grew on level ground, enclosed by hills and mountains on all sides, and yet even in this golden sheltered space their lives were perilous. They were living at the edge of their temperature range, at the mercy of sudden frost.

Giacomo nodded at the hill beyond the trees and explained that you only had to move 200 metres up its side to find conditions untenable for citrus. The Parodi's trees didn't grow in rows, as they would in the large commercial groves of Sicily or Calabria, and even the ancient lemon terraces in Amalfi would have looked ordered in comparison to that lovely fruitful confusion. We walked under a high canopy of ripening fruit, pausing like visitors in a gallery to scrutinize each tree. A pink grapefruit grew next to Limonina, the local variety of lemon; there was Pernambuco, Liguria's version of the Washington Navel orange; a Sicilian *lumia*, or sweet lemon; *cedro ligurese*, the local citron; a lime and several kinds of *mandarini* and *mandararance*, crosses between mandarins and oranges. 'They must cross-pollinate all the time,' I said, and Giacomo answered by showing me a completely spherical lemon: 'It got too friendly with an orange,' he replied.

We stopped by one of the mandarins, a Mapo laden with an unbelievably heavy crop of glistening fruit. Giacomo explained that it was one of the trees planted by his father. 'And until five years ago,' he added, 'it only produced four or five mandarins a year.' One day Giacomo set off with his saw, determined to cut the tree down and graft something more productive on to its rootstock. When his brother saw him he begged Giacomo to give it one more chance, 'and look what happened!' Giacomo said, looking up at the prodigious crop. I was delighted by this story because it coincided exactly with instructions given by Ibn al-Awam in his twelfth-century 'Book of Agriculture'. There's a citrus-related task for virtually every month of the year in Ibn al-Awam's book, but if a tree fails to respond to this intensive level of care there is also a simple cure. 'Let two men carrying an axe approach the tree,' he advises, 'and let the one say, "This tree will be cut down", at which the other should plead for it . . . Then the first must say, "But it bears no fruit", to which the other shall reply, "It will do so this year, and if it does not, then you will be free to do as you please."'[1] That generally did the trick, according to Ibn al-Awam, and he was obviously right.

Every so often we passed a bottle hanging from the branches above our heads. I had seen something similar among citron trees in Calabria. When I asked one of the citron pickers there what the bottles were for, she explained they were used for keeping fish to kill insects that would otherwise damage the fruit. A fish living in a bottle? I struggled with the concept at the time and now I was determined to find out the truth. And of course everyone laughed at me, because the fish is a dead one, an anchovy dropped into a bottle full of ammonia. Giacomo said that wasps, flies and even hornets swarmed to this bait – and there were their smelly little corpses to prove it. However, the absence of sugar in this noxious mixture means that precious, pollinating bees take no interest in it at all.

We worked our way towards the edge of the grove and there we found a group of mature *chinotti*, trees unique to this part of Italy, with a long and very important history in Liguria. Giacomo Parodi explained that the trees had been grown by his father from cuttings taken in the garden of Giorgio Gallesio, one of the great figures in the development of citrus classification and nomenclature in the early nineteenth century. His ground-breaking book, *Traité du Citrus*, was published in 1811. I had already explored the streets of Finale and found the house where he was born, a tall building covered in scaffolding when I saw it. And now Parodi pointed to a villa and terraced garden on the opposite side of the valley and told me that it had been Gallesio's home. I realized I was looking at a garden I had often read about as the open-air laboratory Gallesio filled with different varieties of citrus in order to study their morphology and carry out his carefully controlled experiments in cross-pollination. In *Traité du Citrus* he said, '. . . after close study and thought, I have found great confusion and want of method in their classification . . . I have therefore devoted myself to the close observation of these plants, examining their caprices from their birth to their fruiting and seconding.' From his experiments he deduced that crosses within the same species produced new varieties and crosses between species produced hybrids. Gallesio referred to the *chinotto* as the 'little dwarf from China' and thought it '. . . the most desirable [citrus] variety for ornamenting houses and gardens, being a shrub, dwarfed in all its parts'.[2] His garden is gone, of course, but the dentist who owns the house today has plans to restore it. That's when I'll knock on the door and ask if I can look round.

Gallesio's 'little dwarf' arrived in Italy for the first time through the port at Savona, brought home in about 1500 by a sailor returning from the Far East. Citrus farmers experimented with it and found it grew best on a short stretch of the coast between Varazze and Pietra Ligure. It was here that it mutated and cross-pollinated, evolving into the unique variety identified by Antoine Risso, a Franco-Italian author on citrus in the early nineteenth century, as *Citrus aurantium* var. *amara* subvar. *sinensis*, a name that reveals a genetic link to both the sour and the sweet orange. Today the fruit is commonly known as *Chinotto di Savona* and is unique to the province of Savona. The common name *chinotto*, or *chinottino*, derives from the word *cinese*, or *cinesino*, meaning 'Chinese' or even 'Chinaman', although it's now thought that the tree originated in Vietnam. *Chinotti* are slow-growing and the young trees have a stunted, unattractive look. Their form is indecisive and slightly distorted, as if they were hugging their own chests from sheer embarrassment. If these were the only specimens you had ever seen, you'd ask yourself if *chinotto* was the runt of the citrus litter. And yet the mature trees in the Parodi citrus grove are undeniably elegant. They have narrow branches gracefully held in arching curves, and in spring they bear tight clusters of pure white flowers with a perfume even headier than the *zagara* of sour orange. The tiny fruit hangs in bunches like grapes. The Italians see the unique flavour that its pungent skin and flesh impart to drinks and pastries as one of those things you are destined either to love or to hate.

The *chinotto* harvest always begins in August, when the fruit is still bright green and the unmistakable and breathtakingly bitter flavour of its skin is at its most powerful. Only the largest fruit are picked during the first phase of the harvest and the rest is left on the tree to grow, although *chinotti* are like other people's children – they never seem to get much bigger. They rarely weigh as much as sixty grams and to reach the dimensions of a billiard ball would be the limit of a *chinotto*'s aspirations. As the fruit swells, if that's not an exaggeration, its colour changes first to yellow and then orange, when it is only good for making marmalade.

$\sim$

You might suppose that being very small and very bitter were insurmountable disadvantages for a fruit, but nineteenth-century prints of Savona show the plain between city, mountains and the Letimbro river covered in a regular grid made up of thousands of *chinotti*. Add these to the trees growing on the coast and on stone terraces on the side of valleys leading down to the sea and you realize that important uses had been found for their unprepossessing crop. The fruit is rich in Vitamin C and in a region containing both Genoa and Savona, two of Italy's most important seaports, this was a very significant fact. Sailors knew that citrus cured scurvy long before James Lind confirmed it in 1747, and it was traditional on Ligurian ships to keep *chinotti* on board, like tiny citrus pills for treating the disease. When the British navy made Lind's findings official, Liguria began to supply *chinotti* to the French and American navies. The archives of farming families in Finale Ligure, *chinotto* country, still preserve the invoices relating to this thriving business. *Chinotti* could be stored aboard ship for months on end in a cask full of seawater (a method still used in many Arabic countries to preserve lemons). Each tiny fruit was cut in half, so the brine came into contact with the skin and the flesh, preserving both equally. *Chinotti* ferment very slightly over time when they are stored in this way, which made their skins softer and tempered their sharpness. But a *chinotto* is much more acidic than a lemon and it would always have been a very bitter pill.

If you had visited Savona only thirty years ago, you would have found the same strange object in every bar in the city. It was a tall glass jar decorated in flamboyant, art nouveau style with the moulded figure of a 'Chinaman' holding a parasol. Inside the jar, iridescent-green, candied *chinotti* barely the size of cherry tomatoes floated in a sea of maraschino liqueur. But the glass jars are long gone, and so is the habit of downing a *chinotto* with your espresso on a winter afternoon, although many people still remember the ritual. Take Danilo Pollero, for example, a local agronomist who is now in his forties. 'I know I'm not young any more,' he says, 'because I remember my grandfather taking me into Savona in the 1960s to watch *chinotti* being eaten.' Watch them being eaten? 'Yes, because the fruit was in maraschino, and as a child I wasn't allowed to touch it.' Nevertheless, Danilo can still visualize the barman taking the lid off the jar, dipping in a porcelain ladle and bringing out a single *chinotto* that he tipped into a tiny bowl made from coloured glass. Only the smallest fruit were used for this purpose, so his grandfather could pick one up on a special two-pronged fork with a brightly coloured handle and pop the whole thing into his mouth.

Until thirty years ago these candied fruits were unfailingly served at the end of Christmas dinner in Savona and the western end of the Riviera coastline. They were eaten both for the pleasure of their unique flavour and for their supposed digestive properties. I imagine that the same trays of candied fruit did the rounds during the festive season, being given and received again and again by different people, just as certain bottles of wine and boxes of chocolates are passed around elsewhere. Because in Savona, *chinotti* were the all-purpose present. You'd give them to the doctor you wanted to thank for curing you, as a present when you went to stay with a friend, to your boss, your neighbour or your bank manager.

Go to Savona today and you can sample all of these experiences for yourself at Besio, an elegant bar and *pasticceria* in Piazza Mameli, the main square. Besio serves cakes, drink, sweets and marmalades made from *chinotto* by a company of the same name, and it was here that I sat outside in the November sun and drank Chinotto for the first time. Outside Liguria, or even the province of Savona, few people are familiar with the *chinotto*'s fruit, but the fizzy drink made from a secret combination of herbs, sugar and *chinotti* that takes its name is a different matter. Search for it on a computer and you enter an extraordinary world of fans who are said to have turned Chinotto into something called 'a cult beverage'. Some of them have started websites dwelling on the unique flavour of the drink, which is distinctive because the fruit's irrepressible acidity cuts through the sugar in the recipe to give it a deep and more sophisticated taste than other fizzy drinks. Others are more interested in Chinotto's history and they have reassembled it through old bottle tops, labels and advertising slogans. Few fans seem to know about the fruit that gives their favourite drink its distinctive edge. 'Strange but true,' says one of the websites, 'but *chinotto* is actually a plant.'

When the San Pellegrino drinks company invented Chinotto in 1932, the Fascist government saw it as an ideal home-grown alternative to the American cola drinks that were rapidly gaining popularity. Consequently San Pellegrino's first advertising slogan was a direct challenge to Coca-Cola, and Chinotto was marketed as *L'altro modo di bere scuro*, which could be translated as 'Another dark drink'. Chinotto has always been promoted as a drink for unconventional people, and when San Pellegrino launched a new brand of the drink in 1986 they called it Chinò and promoted it with the slogan *Bevi fuori dal coro*, which might translate best as 'Drink – or think – outside the box'. The words were accompanied by an image of two snakes rising to the music of a snake charmer. The only sign of a third snake was a straw poking out of its basket and into a can of Chinò.

By the time I left Bar Besio dusk had already fallen over Savona, filling the narrow streets and turning them into dark ravines between mountainous buildings. The scent of smoke was everywhere and in the corner of another piazza two men stood among sacks of chestnuts gathered from the woods around the city. They scooped some up off the brazier glowing in an old oil drum and sold them to me for next to nothing. In those quiet streets and broad piazzas it was hard to believe that Savona had been one of the first industrialized cities in Italy. Iron and brass foundries, locomotives and shipbuilding were the dominant industries, but candied fruit, and candied *chinotti* in particular, were also manufactured there on an industrial scale.

When Augusto Besio opened his factory in 1860 it was one of the first factories in town; now Besio is the only candying factory left in Savona. It was the end of November when I paid a visit, and the road outside was blocked by two men unloading a lorry full of chestnuts. The sun was so warm that Christmas seemed unimaginable, but no chestnut would escape the premises until it had been transformed into a *marron glacé* and encased in a gold wrapper. It was a terrible moment to visit a company deeply engaged in festive food. Indoors, the phone rang continuously as businesses placed or chased orders for the candied fruit that would go into Christmas *panettone*, marmalades and Besio's famous *mostarda di Cremona*, mixed fruit preserved in a sweet syrup made from grape must and spices, of which mustard seeds are the most important. It was frantically busy, but that didn't stop Vincenzo Servodio settling down for an hour or two to talk about *chinotti*. Vincenzo, a relation by marriage of Augusto Besio himself, has owned and managed the company for many years.

In the yard outside the factory *chinotti* from last year's harvest floated in barrels of brine that would condition the fruit, making it a tiny bit softer and marginally less bitter. In the old days Besio would have used seawater for this job, but it's not as clean as it once was and now they make up a strong salt-water solution on site. 'The fruit must soak for a minimum of four months,' Servodio told me, 'but a year would be fine.' When they have been removed from their brine bath, the *chinotti* are cleaned and tipped into a turning drum. The slightly abrasive surface acts like sandpaper to remove a thin layer of skin and the glands containing the *chinotto*'s pungent essential oils, reducing a little more the bitterness of the fruit's flavour.

We put on paper suits and hats and went out on to the factory floor. It was sunny outside and, in those great open spaces, light entwined itself with the steam that rose from huge vats of boiling fruit. Figures dressed in white emerged suddenly through a steamy veil and then disappeared again. Peaches, apricots, tiny creamy-yellow pears, dark cherries and mandarins the colour of burnt umber overflowed from buckets and trays on a sunlit metal table. In that stark environment, where everything was painted white or made of stainless steel, the piles of fruit were jewel-like and glistening in the light. 'We have to boil the *chinotti* for a minimum of three hours, because they're so tough,' Servodio explained. 'And then we tip them into a solution of hot water and a 20 per cent concentration of sugar.' And that's the beginning of the candying, the process that, like a culinary version of embalming, will remove all moisture from the fruit and replace it with crystallized sugar, so that each *chinotto* becomes a perfect, mummified copy of itself. Servodio didn't neglect the detail. 'We have to keep the fruit at sixty degrees,' he explained, 'to prevent it fermenting. It stays in the tank for ten days, and as the water evaporates, we add sugar to top it up.' The process is only terminated when the sugar concentration reaches 75 per cent. How even-handed, I thought. Candied, a *chinotto* might rot your teeth, but as a cure for scurvy, it could save them.

At last it was time to taste a candied *chinotto* in syrup, a double dose of sweetness for someone who doesn't appreciate sweetness much at all. 'Well, here goes,' I said, prodding the thing with a toothpick and putting it whole into my mouth. Then the magic began, the unexpected magic of a first *chinotto*. Its skin was still slightly hard, but as I bit through it to the inside, the flesh was meltingly soft. And with that bite came the intense flavour, the sweetness of the candying alleviating but failing to mask the bitterness of the fruit. And that bittersweet taste, which is the essence of all the oranges you've ever known, leaves a sensation almost like fizzing in the mouth long after it's swallowed. And even after that's gone, it infuses your breath with its perfume. I was a convert. 'Can I have another?'

~

The popularity of candied fruit began to wane after the Second World War, and although several different companies manufactured Chinotto by this time, the extraordinary distribution powers of cola-producing multinationals eventually drove all but a very few of them out of business. Coupled with this drop in demand came the unusually hard winter of 1956, when the peril of cultivating citrus in the marginal conditions of the Riviera was proved once and for all, and severe frosts killed almost all of Liguria's *chinotto* trees. The *chinotti* used for drink manufacture today are *Citrus aurantium* var. *myrtifolia*, the original and less special variety of *chinotto* that grows in southern Italy and Sicily. Danilo Pollero, who has made himself a determined champion of the *chinotto* cause, can add another layer to the sad story of its demise in Liguria. 'If a *contadino* loses a valuable tree,' he told me, 'he'll replant it, but if the tree was earning him nothing in the first place, he won't bother.' And then Danilo got to the heart of the matter. 'Even though candying was an important industry in Savona, only two or three companies were involved, and they reached agreement among themselves about the price of *chinotti*.' They conspired to drive prices down, and the fruit was soon worth so little that nobody could be bothered to grow it any more.

At this point the *Chinotto di Savona* could have been forgotten, but at the end of the 1990s, when the city's traditional industries were in crisis and 70,000 people had lost their jobs, Savona reinvented itself. The famous docks, once a centre for shipbuilding and citrus export, were transformed into a tourist port for cruise ships and their passengers. This new focus on tourism inspired the city council to look for a local product to use as a symbol of the city. Meeting followed meeting, years passed, and yet no one could identify a perfect product for the role. One day Danilo Pollero, who works for the local council, happened to be in the city hall when the council officer in charge of the project burst out of yet another meeting. He overheard enough to realize that no solution had been found and he quickly proposed the *chinotto* as the symbol of Savona. But unfortunately, a *chinotto* census taken in 2003 and 2004 revealed only 118 trees growing in Liguria, and so hundreds of new trees had to be planted. In 2004 the *chinotto* attracted the attention of Carlo Petrini, founder of Slow Food, because it had all the qualifications necessary to be nominated for a *presidio*, one of the small bodies set up within the movement to protect traditional products in danger of extinction. This brought it to the attention of a wider public, including the citizens of Savona, who seemed generally unaware of its existence. Today *chinotti* fuel a number of tiny local industries producing delicious marmalades, liqueurs, *amari* and *digestivi* that are sold on market stalls and in shops specializing in local foods. And as well as being candied, the *chinotto*'s peel is used by Besio to perk up their special *amaretti* biscuits and Christmas *panettone*.

~

The *chinotto* was just one element in an industry that also encompassed lemons, mandarins, oranges and citrons growing in Liguria, and can be traced back to the fourteenth century, when Sanremo shipped 50,000 oranges to Avignon[3] in a single year, and exchanged lemons for grain with Arles.[4] By the seventeenth century the extraordinary volume of citrus in Liguria perfumed the air itself, so that Giovanni Battista Ferrari in his book *Hesperides* (1646) rebaptized the Gulf of Genoa 'the perfumed sea'. English diarist John Evelyn made the crossing from Cannes to Genoa in October 1644, when his ship was hit by a storm so severe that an Irish bishop and a priest on board unnerved other passengers by giving each other the last rites. Perhaps this made Evelyn especially appreciative of the sight of 'pleasant Villas and fragrant orchards', and an offshore wind carrying 'the peculiar joys of Italy, in the natural perfumes of Orange, Citron and Jassmine flowers, for divers leagues to seaward'.[5] Dickens made no mention of this experience when he approached Genoa in the same way in 1844, although after his arrival he did dwell on the odour of the streets, '. . . a peculiar fragrance, like the smell of very bad cheese, kept in very hot blankets'.[6]

In the most sheltered areas of the Riviera citrus grew for centuries in gardens and open terraces cut into steep hills above the sea. Dickens described 'watching as it gradually developed its splendid amphitheatre, terrace rising above terrace, garden above garden, palace above palace, height upon height . . .',[7] and he never forgot Strada Nuova, the main street, and '. . . the terrace gardens between house and house, with green arches of the vine, and groves of orange trees, and blushing oleander in full bloom, twenty, thirty, forty feet above the street . . .'[8]

The most famous garden in the city belonged to Andrea Doria, an admiral in Emperor Charles V's navy, who had his palace and grounds carefully designed to overlook every inch of the harbour. John Evelyn described his garden as a sweep of terraces reaching 'from the very Sea to the Summit of the Mountaines', and noted that some of the terraces were 'full of Orange-trees, Citrons, Pomegranads, Fountaines, Grotts and Statues'.[9] Thankfully, the lovely view over fountains, terraces and the lush canopies of citrus trees to Palazzo Doria and the sea was immortalized by the Dutch painter

Jan Massys when he made them a backdrop to the alert and voluptuous figure of his *Cytherian Venus* in 1561. I should have known when I went to Genoa in search of Doria's garden over twenty years ago that it was long gone, but I still remember sitting in a bleak car park overlooking the flyover that cuts Genoa off from the sea and noticing a sixteenth-century fountain among the parked cars. Eventually I realized I had parked on ground that had once been the uppermost terrace of the garden. Since then Doria's house, now known as Villa del Principe, has been restored and opened to the public. The flyover overshadows the surviving garden immediately in front of it, which has been replanted with species commonly used in the sixteenth century. Pots of lemons stand in a circle around the enormous Neptune fountain at its centre, all of them labelled with a bleak warning not to eat their fruit 'because it has been sprayed with poisons'.

Citrus is still a common sight throughout the year in Genoa's gardens and courtyards, and when the Ligurian poet Eugenio Montale glimpsed pots of lemons from twentieth-century streets made mean and ugly by winter, it seems likely he was writing about his native city. In the final verse of 'I Limoni' the sight of lemons through an open door transforms the dismal urban scene. The impact on him, on us, is instantaneous:

> e il gelo del cuore si sfa,
> e in petto ci scrosciano
> le loro canzoni
> le trombe d'oro della solarità.[10]

> (melting our frozen hearts
> and filling our souls with their song,
> a glorious trumpeting,
> a golden roar of sunlight.)

While citrus disappeared from the gardens of Sicily's villas and palaces as soon as it began to be cultivated as a commercial crop, the aristocracy in Liguria had none of the Sicilians' snobbish qualms about selling the fruit from their ornamental garden trees. This habit shocked Arthur Young, an Englishman who made an agricultural tour of France and Italy towards the end of the eighteenth century. He had already been warned about gardens on the Riviera that were 'a mixture of half garden and half farm', and yet he seemed utterly unprepared for what he saw. 'Thus the garden,' he remarked afterwards, 'which with us is an object of pleasure, is here one of economy and income, circumstances that are incompatible. It is like a well-furnished room in a man's house which he lets to a lodger.' He was equally upset by the notion that you couldn't go out into the garden and casually pick fruit for your own enjoyment, as a 'certain momentary and careless consumption is a part of the convenience and agreeableness of a garden; a system which thus constrains the consumption destroys all the pleasure'.[11] Young was a vigorous advocate of improvements in agriculture and I wonder what he would have thought of a traditional treatment of citrus pests, still in use in 1872. An account of it can be found in a letter written by Ludwig Winter, a German gardener who worked for Englishman Sir Thomas Hanbury at La Mortola, near Ventimiglia, where Hanbury and his brother Daniel amassed one of the most famous plant collections in the world. Winter was fascinated by local traditions and in July 1872 he wrote to Daniel describing a method used to control a beetle with a voracious appetite for lemon blossom. Apparently a monk from Ventimiglia was thought able to destroy all kinds of pests simply by blessing the trees and cursing the insects that attacked them. Winter described this ceremony, the handsome fee paid to the priest and the splendid dinner offered to him by local farmers, also remarking that 'the beetles seem to be quite insensible for these insults of their chivalrous race and make now their appearance in far greater number as any time before'.[12]

The combination of beauty and utility that so shocked Young delighted Tobias Smollett. Standing on a rampart on the edge of Nice, or Nizza as it was then, he said, 'The small extent of the country which I see, is all cultivated like a garden. Indeed the plain presents nothing but gardens, full of green trees, loaded with oranges, lemons, citrons and bergamots, which make a delightful appearance . . . roses, carnations, ranunculus, anemones, and daffodils, blowing in full glory, with such beauty, vigour and perfume, as no flower in England ever exhibited.'[13] If Smollett had moved along the coast as far as Bordighera and Sanremo, he might have been surprised by the rare and exotic sight of date palms, *Phoenix dactylifera*, growing among citrus trees. Locally it is said that St Ampelio brought palm seeds to the Riviera from Egypt in the fourth century and planted them around the natural harbour below Bordighera. Others believe that the Dominicans brought the first palms to Liguria in the sixteenth century. The warm climate and sandy soil provide ideal conditions for palm trees and the skyline in Bordighera is still dominated by their ragged outlines. The palms are the legacy of a thriving industry that was once intimately linked to the cultivation of citrons on the Riviera. In Bordighera they were grown in gardens and on terraces in the Valle di Sasso, and their fronds were harvested twice a year. A late summer harvest was timed to supply palm fronds for Sukkoth, the Jewish Festival of Tabernacles, which takes place in early autumn. When Jews enter the tabernacle during Sukkoth they must take a number of things with them, including a palm frond and a citron. Farmers on the Riviera built up a lucrative business shipping palm fronds and citrons out of Genoa and Savona to Jewish communities all over northern Europe. A decree in Sanremo in 1664 set the price of citrons for the Jewish market.[14] Earlier in the year fronds from the spring harvest were sent to the Vatican in Rome in time for Palm Sunday (the Sunday before Easter), along with *palmureli*, woven palm-leaf ornaments rather similar to the traditional English corn dolly. When the palm trees reached a certain height their leaves were bound in sacking. This was a way of blanching the leaves, giving them a unique silver colour.

In Sicily citrus cultivation, with its combination of high risks and extraordinary profits, was unregulated, while in Liguria every aspect of the industry was tightly controlled by law. The fruit was picked by public employees working to strict standards established by decree in Menton, Sanremo and other Riviera towns. Both fruit and the juice extracted from it were inspected or tested to ensure the high standard of products for export, and local councils set prices and levied taxes. This thriving industry encompassed trees as well as fruit. In Nervi, a small town just outside Genoa, it was too windy to grow citrus in open ground but nurseries were established there to cultivate pot-grown citrus trees for export to northern Europe.[15]

A book published in 1692 describes Genoese merchants taking bare-rooted orange and lemon saplings to Dutch markets in March, April and May each year.[16] Young saplings grown in Nervi and on narrow terraces above Finale Ligure and Savona were sent across the Alps to Austria and Germany in cases filled with earth, because it was thought the trees would be more likely to grow and prosper in familiar soil. Mules were the only animals sure-footed enough to transport the heavy cases across the mountains, and although exceptionally strong, these mules were also vicious. According to Tobias Smollett, they 'have such an aversion to horses, that they will attack them with incredible fury, so as even to tear them and their riders in pieces'. Smollett's advice to riders glimpsing a mule train was to gallop away in the opposite direction.[17]

∼

The remarkable beauty of the Riviera and the mild winter climate attracted two very different groups of visitors. The first were artists, dazzled and entranced by the extraordinary clarity of the light. When Monet came to Bordighera from northern France in 1884, he is said to have found it difficult to paint lemon and orange trees silhouetted against the sea, and yet his painting entitled *Sous les Citronniers* captures perfectly the vibrant, dusky light beneath laden trees in an orange grove. In a letter to Alice Hoschedé he described with delight the wild profusion of almond and peach trees all muddled up with palms and lemons in a garden.[18] The largest garden belonged to a Signor Morena. Rather like Andrea Doria's sixteenth-century garden in Genoa, it stretched from the top of the hill in

Bordighera to the sea. When Monet first visited Signor Morena to ask if he could set up his easel in the garden, he returned to his hotel with armfuls of flowers, oranges, mandarins and sweet lemons. 'This garden is unlike any other,' he wrote in a letter; 'it's a dream, where all the plants in the universe grow naturally.' Morena's villa is now called Villa Mariani, after the painter Pompeo Mariani, who lived there from 1909 until his death in 1927. It is possible to visit it and see the views that Monet painted, although the garden no longer stretches down the hillside to the sea.

The Riviera's mild winter climate also attracted invalids and in particular sufferers from tuberculosis. This trend began in the late 1850s when Dr James Bennett, an English gynaecologist suffering from TB, came to Menton for his own health. When he recovered he wrote a book that was published in 1861 to promote the Riviera as an overwintering station.[19] 'I believe that the time is fast approaching,' he said, 'when tens of thousands from the north of Europe will adopt the habits of the swallow, and transform every town and village on its coast into sunny winter retreats.' How right he was, and citrus trees were a vital element of the vivid impression made by the Riviera upon its frail visitors.

The romance of the Riviera was greatly enhanced in 1855, when Giovanni Ruffini published a novel in English called *Doctor Antonio: A Tale of Italy*. The book begins as a story about the relationship between Lucy Davenne, a beautiful young English girl injured in a carriage accident on the road between Nice and Genoa, and Dr Antonio, a political exile from the Bourbon regime in Sicily. The 'rich odour of orange and lemon trees'[20] surrounds the doctor and his young patient, creating a background for their unspoken romance. When Dr Antonio allows Lucy to leave the house for the first time after her accident, he leads her into the garden, where the ground is covered by 'a thick layer of orange and lemon blossoms, out of which came in strong relief a profusion of violently red wild poppies'.[21]

When Lucy returns after many years to the old house, she finds the garden 'a perfect wilderness of weeds and brambles; the once luxuriant grove of lemons and orange trees has dwindled into a scanty assemblage of shivered, scattered, skeleton-looking trunks, – the few dry reddish leaves, still hanging on the branches, look as if they had been scorched by lightning . . .'[22] This is the Riviera without Dr Antonio, the Riviera without romance, and the scene marks a change in the tenor of the book. Ruffini himself was a political exile from Sicily and at this point his novel evolves into a highly politicized account of the end of Bourbon rule on the island. The story of Lucy and Dr Antonio, which had been so predictable and seemed destined to end in marriage, takes an unexpected turn when Dr Antonio favours his political ideals over personal happiness. Yet it was the image of the Riviera in the first part of the book, carpeted in citrus trees and wild flowers, that seemed to endure, and the book was enormously popular.

Unification came only five years after *Doctor Antonio* was published, and in 1860 a referendum was held, allowing the citizens of Mentone and Nizza (now Menton and Nice) at the eastern end of the Riviera to vote on whether they wanted to stay in Liguria and become part of a united Italy, or join the French Republic. Nizza had already been promised to Napoleon, which was awkward as it was Garibaldi's hometown. When the inhabitants voted to join France, it's thought their decision was inspired partly by a belief that lemons would find a better market there.

After 1860 the French and Italian Riviera developed slightly different characters. The coast on the Italian side of the border was never quite as elegant as the French. As Edward Lear put it in a letter to Lady Fortescue written from Sanremo in 1872, 'there ain't a creature here you would know, I think . . . we are all humdrum middleclass coves & covesses, & no swells'.[23] Nevertheless, for those coves and covesses the presence of orange, lemon and citron trees in Liguria was a potent element in the romance of the Mediterranean.

The Sweet Scent of Zagara

Throughout the nineteenth century essential oils from the flowers, leaves and skin of both sour orange and *chinotto* trees grown in the province of Savona were used by perfume manufacturers in Grasse to increase the strength and persistence of fragrances that might be thought to encapsulate the romance and beauty of the Riviera. The short *zagara* harvest took place in April and May, when the flowers were picked early in the morning, just as they began to open. They were only harvested on sunny mornings, because damp or overcast weather affected the scent of their oil. If they were picked too soon the oil yield was low and their immaturity produced an unwanted 'green' note in the oil's fragrance. Too late, and much of the precious volatile oil would evaporate while the blossom was being transported to the stills. *Zagara* from trees grown in Liguria used to be packed in barrels between layers of salt to preserve it during the journey to Grasse.

A scent as evocative, sensual and suggestive as the essential oil extracted from citrus blossom is a gift to the perfume industry. *Chinotto di Savona* was virtually extinct a decade ago, but becoming a symbol of the city has brought it back into the limelight, and the fruit's connection with the perfume industry has been renewed by Marco Abaton, a parfumier in Savona. He has combined all the different notes of the *chinotto* in a new fragrance he calls Il Chinotto in Fiore. Abaton set out to make a perfume that would recreate the unique sensation of drinking Chinotto or eating the candied fruit. He found it essential to consume both of these things during the design process because he wanted his perfume to reproduce the very particular combination of the tart notes of the fruit with the sweetness of sugar. When it first escapes from the bottle, Il Chinotto in Fiore is like a blast of incense, a scent that is spicy

and redolent of the cool, empty spaces of a vast cathedral. This is what Marco describes as the woody note, derived from the leaves and bark of the tree. But it alters almost immediately, opening up to release the sharp citrus notes of the essence of *chinotto* extracted from the peel by infusion. It's hard to keep up with the perfume's rapid changes on my wrist. Now it's softening, becoming warmer and more floral as the sweetness of the *chinotto*'s flower breaks through. Abaton has done what he set out to do: he has recreated in scent the multidimensional and surprising experience of eating a candied *chinotto*.

You only have to fly into Catania airport in Sicily in late spring to appreciate the raw power of *zagara*, for as the doors open, the scent will bludgeon its way on to the plane. The smell has many layers. Gabriele d'Annunzio, the nationalist writer whose explicit poetry and prose scandalized Europe in the late nineteenth and early twentieth centuries, described the scent as 'candid, unripe, infant-like'. And it's true that at its most superficial *zagara* is a bright, carefree perfume, the kind an adolescent might wear to her first party. But it has depth and below its innocence is something cloying and almost fetid, like the stale air in an elderly actress's dressing room.

In spring, green buds form like a haze all over the sour orange tree, opening into pure white, five-petalled stars around a clutch of yellow stamens. *Zagara* fills the air with an invasive, migrant scent, a scent heavy enough to roll across distance, so that you often smell it in Sicily without seeing the trees themselves. Giuseppe di Lampedusa uses this dense, suggestive perfume in *The Leopard*, where it saturates his vision of nineteenth-century Sicily with decadence and sensuality. In the garden of the Prince of Salina's palace in Palermo, 'an erotic waft of early orange blossom' fills the air, even though the prince can't see the orange grove beyond the garden wall.[1] And the scent is there again when he sets off on a 'low love adventure', driving into the centre of Palermo in a stuffy carriage and attempting to disguise the purpose of his outing by taking the family priest with him. The orange trees are concealed once more, this time by the gathering dusk, and yet their scent infiltrates the scene, heightening the prince's erotic expectations. 'Now the road was crossing orange groves in flower, and the nuptial scent of the blossoms absorbed the rest as a full moon does a landscape; the smell of sweating horses, the smell of leather from the carriage upholstery, the smell of Prince and the smell of Jesuit, were all cancelled out by that Islamic perfume evoking houris and fleshly joys beyond the grave.'[2]

The essential oil of sour oranges began to be known as neroli in the late seventeenth century, when it was named after Anna Maria de la Trémoïlle, Princess of Nerola. Like many of her contemporaries, the princess liked its scent so much that she used it to perfume both her gloves and her bathwater. The French princess took her title from her second marriage in 1675 to Flavio Orsini, Duke of Bracciano and Prince of Nerola, a small town near Rome. She was twenty-two years younger than her husband and after his death in 1698 the bittersweet, slightly metallic scent of neroli became the scent of intrigue as she immersed herself in European political life. She gathered supporters for the French cause in the War of the Spanish Succession around her, and actively promoted marriage between the thirteen-year-old daughter of the Duke of Savoy, Maria Luisa, and Philippe V, the French king of Spain. Anna Maria now called herself the Princesse des Ursins, a witty adaptation of the Orsini name stripped from her after the death of her husband, and when she accompanied the child queen to Spain to become head of her household, the scent of neroli crossed another frontier.

Neroli is still the principal ingredient in over 12 per cent of modern perfumes. It is very expensive, but petitgrain, which is extracted from the young shoots, twigs and fresh leaves of sour orange trees, is a lower-grade, less valuable oil. Its French name, 'small grain', refers to the original source – green, immature fruit so small that they resembled grain. Cheaper petitgrain oil is sometimes mixed with neroli, but this gives it a bitter, woody scent. Petitgrain, neroli and essential oil expressed from the rind of the fruit all have slightly different qualities, but overall their presence in a perfume imparts bright, fresh, tonic notes.

Dogged Madness

Limonaie *on Lake Garda*

⚬⚬⚬

Foreign visitors traditionally associate the scent of citrus blossom with the Mediterranean and they have never expected to find it in the far north of Italy. And yet when Goethe arranged to be rowed across Lake Garda in Lombardy during the summer of 1769, he looked back towards the village of Limone on its west shore and saw 'terraced hillside gardens planted with lemon trees, which made them look at once neat and lush'. Goethe had already noted that his journal entry was 'Written on a latitude of 45° 50''', but despite its northerly position, the place had all the romance and relaxed charm that Goethe habitually associated with the Mediterranean. 'The people lead the careless lives of a fool's paradise,' he remarked. 'To begin with the doors have no locks, though the innkeeper assures me that I would not have to worry if all my belongings were made of diamonds. Then the windows are closed with oil paper instead of glass.' He was a little less enamoured of the lifestyle when he discovered that 'a highly necessary convenience is lacking' and was invited by a servant to make use of the far end of a courtyard. 'Where?' he asks. 'Anywhere you like,' the servant replies.[1]

The thriving citrus business that Goethe glimpsed in that unusual context is long gone, but the western shore of the lake still retains its unmistakable imprint. Lemons grown in these cooler climates are particularly bitter, a characteristic that seemed to make them popular in northern Europe, but to cultivate such a high-risk crop in the challenging conditions of northern Italy demanded a degree of dogged madness that renders the whole subject of citrus production in these places all the more intriguing.

~

Lake Garda, about halfway between Venice and Milan, is surrounded by the snowy peaks of the pre-Alps. It seems an unlikely spot for lemon farming, and yet lemons grown on the lake shore were once the basis of a thriving export industry that took fruit all over northern Europe. Go to Garda today and you will find traces of the unique architecture of this particular form of citrus farming all over the lake's west shore. The route takes you down the busy motorway linking Milan and Venice, a road I often travel when I'm leading a garden tour in the Veneto. Sitting high in the front of a coach, I'm scarcely aware of the juggernauts that work this road from end to end. It was a very different matter moving among those huge beasts when I drove to Garda myself a couple of years ago. While still on the motorway I realized I was on the old route that once carried lemons from Lake Garda to Germany, Austria, Hungary, Poland and even Russia, and so when lorries hogged the middle lane, stepped on my heels and sounded their bellowing horns in my wake, I told myself there was a crowd on the Citrus Route that day.

The road from the motorway dips down into Desenzano and then to a much narrower road running between the mountains and the west shore of the lake. On that day in early September I had set off from Venice in heat better suited to mid-July, but as I drove along the shore tumultuous clouds began to form, and the lake water gathered in choppy waves. Usually the sloping ground on the west shore is bathed in sun, and it's always sheltered by hills from some of the cold winds off the Alps. Summer heat is trapped and stored in the vast volume of water, and in winter this warmth is gently released, creating a peculiarly mild microclimate for this northerly latitude. Citrus trees have been cultivated beside the lake ever since the thirteenth century, when Franciscan friars grew them for the first time in the precincts of their monastery in Gargnano, about halfway up the lake's west shore. It is still possible to visit their lovely fourteenth-century cloister in the centre of town, and see the different combinations of carved lemons, leaves and flowers that decorate each column.

Citrus farming was always a gamble in these parts, just as it was on the Riviera, because every decade or so the night-time temperature would plummet, sometimes dipping as low as minus eight degrees Celsius, and then the citrus crop and even the trees themselves would be in grave danger. You might think this a good enough reason to abandon any plans for growing citrus on a grand scale, but at the end of the eighteenth century the Bettoni, who still own a lovely villa on the water's edge in Bogliaco, began to build the first of the extraordinary lemon houses that soon covered the sloping landscape between Gargnano and the small town of Limone. I saw one for the first time many years ago, when I was out on the lake in a boat. I had no idea then what the enormous pale pillars were, rising so grandly between the mountains and the shore. Since then I've seen several lemon houses and I know that each one consists of several deep terraces (*còle* in the local dialect) linked by stone steps and lined with square pillars nearly ten metres high. Large chestnut rafters called *sparadòs* run between the tops of the pillars and in winter they are used to support wide planks (*às*) that create a sheltering roof for the trees. Smaller beams (*canter*) are attached horizontally to support narrower planks (*mesì*) for the walls, glass panels (*envédriàe*) and doors (*üsére*). All of these panels and roofing planks are removed and stored during summer in rooms called *casèl* at the ends of the terraces, and then the trees grow in the open air, sheltered only by the thick walls that surround the garden on three sides, leaving it open east-south-east.

You might suppose that the small town of Limone was named for the fruit that grew all around the village, and yet lemons were not cultivated there on a grand scale before the eighteenth century and the name actually derives from the Latin word *limen*, meaning 'border' or 'boundary', a reference to its position close to the frontier of Roman rule. It's easy to drive up the west shore of the lake today, but when the Bettoni wanted to build their first lemon gardens, there was no road and all of the materials had to be loaded on to boats and carried by water. Until then the inhabitants of Limone had scratched a living from fishing in the lake and growing olives, but now there were jobs to be had as manual labourers, builders, carpenters and lemon gardeners. All this activity attracted attention outside the village and soon entire families began to migrate to Limone in search of work. The Bettoni were in business, but they didn't want their family name sullied by vulgar associations, and so they called their company Bentotti, which won't have fooled anyone. An archive containing diaries, account books and letters is still housed in their lovely villa at Bogliaco. Most of the letters were received in the mid-nineteenth century from customers in Prague, Krakow, Vienna, Warsaw, Konstanz, Lviv, Trieste and Milan, places where the exceptionally bitter juice and aromatic peel of the Garda lemons were especially appreciated. Bentotti's northern European customers would pay two or three times as much for Garda lemons as they would for the less acidic varieties of lemon grown in Sicily or Calabria.

Limone was the first village to cultivate lemons in the new lemon houses, but other villages on the western shore of the lake soon joined in the lucrative trade. All of them had a geographical advantage over lemon producers in the south of Italy because the journey from the lake to the shops of northern Europe was much shorter, and consequently their lemons were bound to arrive at their destination in better condition. Lemons for export were graded using a series of rings of different sizes. The finest fruit was destined for Hungary, Poland and Russia. It was carefully wrapped by a woman-only team and packed in heavy-duty wooden cases. Protected in this way, the lemons would last for up to six months. Fruit of an inferior quality was sent in lightweight boxes and without wrappers to Milan, Verona, Brescia, Parma and Modena. There was no road to Limone until 1931. Until then there were only mule tracks across the mountain's flank, and they were tiny, tortuous and entirely unsuitable for transporting a huge volume of fragile fruit that would deteriorate rapidly if it was bruised. Instead the fruit was loaded on to boats and sent either to Torbole at the lake's far end, from where it would travel north via Bolzano, or in the opposite direction to the market at Desenzano. Only the poorest fruit was retained for local consumption.

~

On my September visit to Garda the storm broke as I reached Limone. I walked through narrow streets full of baffled German tourists robbed of late-summer sun. I blundered out of the rain and into the back of an *osteria*. There was a tourist menu and a smell of frying *schnitzel*, and yet my *spaghetti alla limonese* was truly delicious. You might think I'd tire of pasta on a citrus theme, but this combination of anchovies, black olives, parsley and the grated peel of lemons was very fine. I had only one reservation. The menu listed the ingredients of each dish. Mine was supposed to be made with Garda lemons, and yet I knew that those bitter little fruit were virtually extinct, and it seemed unlikely that the crop from any surviving trees would be supplying the kitchen of a tourist restaurant.

After supper I returned to my hotel above the lake and all night a hot gale blew, filling the room with the sound of breaking waves. In the morning when the town was tranquil and rinsed clean I set out to meet Domenico Fava, a local man whose ancestors came to Limone to work for the Bettoni. Fava has spent thirty years accumulating information about Limone's lemon industry. He is an habitué of the Bentotti archive and has written his own account of the history of the local lemon industry and built up a collection of thousands of postcards showing views of the town during the nineteenth and early twentieth centuries.[2] He led me through the pretty, tourist-clogged streets and paused beside a high wall. There were two small holes at head height and when I put my eye to one of them I found I was looking into a lemon house. The wall was enormously high and, because of the hole through it, I could see it was about half a metre thick. Inside was a sunny sheltered space enclosed on three sides. The high back wall protected it from cold wind off the mountains and its south side was open to the lake. Domenico explained that the hole in the wall would originally have been made to accommodate a vine. It was common to plant other crops, like grapes, capers or different kinds of citrus fruits, on the warm south walls of a lemon house, but vines were trained through the wall to grow outside the house so they didn't shade the trees. Both vines and trees were long gone and the garden contained nothing but grass and pillars that strode across my view of the lake.

We saw numerous other lemon houses as we walked through Limone, their pillars standing about the town like the skeletons of gigantic prehistoric beasts, bleached bones stark against the dark backdrop of the mountains. Either they lay empty or they had been converted into swimming pools, built into the structure of hotels or made into houses. Up on the mountainside Domenico led me to one of the enormous water tanks used to store snow melt and spring water for irrigating the trees. The abundance of water, from the mountains or the lake itself, was another advantage that Garda's lemon farmers originally had over their Sicilian counterparts. Far below us on the water's edge we could see the skeleton of Reamòl, one of the first gardens ever built by the Bettoni – and what's more, it was built by Domenico's ancestors. Family names are perpetuated in the structure of the *limonaie* and at Reamòl one of the terraces, or *còle*, was called *còla Fava*.

At the beginning of the nineteenth century Garda's thriving lemon trade started to be dominated by unscrupulous middlemen who creamed off the profits of the producers by taking over all their negotiations with citrus merchants, shopkeepers and the German pedlars who came specially to Lake Garda to buy lemons. In 1840 an agricultural cooperative called the Società del Lago di Garda was set up in the old Franciscan friary in Gargnano. A history of the Società written in 1883 conveys a vivid impression of the situation it aimed to resolve:

> An enormous desire for gain simmered among the merchants and the middlemen, who imposed their terms on the growers. A tip or a promise was enough to devalue fruit of one grower, and boost the price of another's . . . So those noble fruits . . . were in the hands of greedy speculators who did not recognize anything but their own profit, obtained by any means, by any sacrifice of their fellow countrymen.[3]

The Società sorted out the speculators and crooks, but within fifteen years of its foundation Garda's lemons had been infected by gummosis. For twenty years gummosis ravaged the lemon gardens, and production fell from 10 million lemons per year to just 3 million. The situation worsened in 1880, when two exceptional floods destroyed or damaged many of the trees on the shoreline, sending the Garda lemon industry into its final decline.

~

D. H. Lawrence spent the winter of 1912–13 in Gargnano and wrote a very accurate account of the dying days of lemon cultivation in *Twilight in Italy*. To his eye the naked pillars of the lemon houses were 'like ruins of temples . . . as if they remained from some great race that had once worshipped here'. As snow began to fall over the mountains in November he saw the space between the pillars 'blocked in . . . with old, dark wood in roughly made panels. And here and there, at irregular intervals, was a panel of glass, pane overlapping pane, in the long strip of narrow window.' He watched in awe as two men clambered about the fragile lattice of chestnut poles high above the ground, 'talking and singing as they walked across perilously . . . In their clumsy zoccoli [clogs] they strode easily across, though they had twenty or thirty feet to fall if they slipped.' He described the lemon houses in winter as 'blind, dark, sordid looking places' that were cold and dark inside, 'almost like being under the sea'. These conditions threw the trees into a semi-dormant state, making them less vulnerable to fluctuations of temperature and delaying the ripening of the fruit until early summer, when it would fetch the best price. On warm days the glazed panels on the front of the houses had to be opened to just the right degree to adjust the temperature inside and prevent the air becoming too humid. Lawrence described lying in bed on a sunny winter morning and listening to the 'little slotting noise which tells me they are opening the lemon gardens, a long panel here and there, a long slot of darkness between the brown wood and the green stripes'.[4]

During the First World War, everyone living on the shores of Lake Garda was forced to abandon their lemon gardens and evacuate the area. And to make matters worse, much of the wood normally used for enclosing the lemon houses was requisitioned for use as shuttering in the trenches on the front line. As a result many lemon gardeners were forced to leave their trees unprotected after the war and they lost most of them during the exceptionally cold winter of 1928.

In 1932 any surviving lemon growers were given fresh hope when the new Gardesana road was built along the west shore of the lake, linking Limone for the first time to Gargnano. This brought tourists to Limone and photographs taken between the 1930s and 1960s show them buying lemons from colourful roadside stalls. But the tourist trade wasn't lucrative enough to keep such a labour-intensive system of lemon production alive and the last few lemon gardens were eventually abandoned. Now, however, the town council in Limone has restored and replanted two lemon gardens that are open to the public. In Gargnano there is another *sardì* in perfect working order. It belongs to Giuseppe Gandossi, who bought it forty years ago. It was the house that he and his wife fell for, but it just happened to have an abandoned lemon garden attached to it. There was planning permission for another house on the lemon terraces, but Gandossi had no intention of building there and so he decided to restore the *sardì*. He worked full-time in those days, running the men's outfitters that he still owns in Gargnano, and so he already knew that the enormous tasks of restoring and managing the lemon house could only be attended to in his spare time.

The garden had been abandoned for twelve years by the time the Gandossi bought it, and only six out of the original 175 trees survived. Almost as soon as contracts were exchanged a terrible frost destroyed all of the remaining trees. Fortunately Gandossi knew someone who was dismantling an old lemon garden elsewhere on the lake, and he was able to buy a few Garda lemon trees from him. And so it was in Gandossi's garden that I saw a genuine Garda lemon for the first time. It wasn't much to look at, small, smooth-skinned and perfectly oval, but after I got home I squeezed its juice over a plate of insipid risotto and the transformation was instant. It lifted and strengthened the flavours, creating impact in a vacuum. It had the unique and scented bitterness of the Garda lemon and, as Gandossi said, 'No one likes a sour orange, but in a lemon, bitterness is prized.'

~

Nothing can prepare you for the surprise of seeing a Garda lemon garden for the first time. The size strikes you at once: a cathedral with an open roof, a high-rise block without floors. Each tree grows in its own plot or *campàa*, a twenty-square-metre patch of ground between two sets of pillars. The trees are allowed to grow to eight or nine metres tall. They put down shallow, horizontal roots and this makes them rather unstable, but it doesn't matter because each one grows inside a supportive scaffolding of chestnut poles. We clambered all over the lemon house with Gandossi. He showed me the irrigation channels that carry water along the back walls of the terraces, took me into the storerooms at the end of each terrace, where the bleached boards and glazed panels that clothe the structure in winter were stacked, and into his workshop. Gandossi cultivates his trees in a strictly traditional manner, employing tools and utensils used for hundreds of years in these parts. You can only find these things in museums today, so if he wanted any himself, he had to make them. In the workshop he showed me the leather buckets made from donkey skin that were always used by lemon pickers. And then there were the buckets full of wooden nails for securing the roofing planks in winter, which he whittled in the evenings after work.

We ended up at roof level on the top terrace, where, when I could tear my eyes away from the enormous drop beneath our feet, I began to grasp the labour involved in cladding that vast structure with boards and panels every autumn, and to understand the dogged madness required to grow lemons at 46° North. It used to take Gandossi a fortnight of perilous hard labour to shift all the cladding out of the storerooms, to scramble across the skeleton of pine raft-ers and put the roofing planks in place, to secure them with his handmade wooden nails, and shutter the sides of the terraces with boards and glazed panels. Now he's a little older, his son helps him out, but the winter work doesn't end there, because each crack in this vast, temporary structure must be stuffed with dried grass to insulate it against draughts. And once all that work is done, Gandossi still has to stay on red alert in case of frost. Traditionally, lemon gar-deners kept a bowl of water on their bedroom windowsills. On cold nights, they would feel its consistency with their fingertips. Any hint of thickening or freezing would send them into a frenzy of ac-tivity, because fires had to be lit at the end of each terrace and tended throughout the night to warm the air and prevent the frost destroy-ing the fruit and trees. In the nineteenth century this crude and rather hazardous system was replaced by stoves and heating pipes.

GANDOSSI'S LIMONCELLO

To make a litre of limoncello Gandossi uses twenty-four lemons. Ideally they are completely fresh and slightly green. He peels them thinly, without the pith, and puts the peel into an airtight container with the contents of a litre bottle of 95 per cent alcohol that he's picked up for next to nothing in the supermarket. Some people advise leaving the peel to steep for a couple of weeks, but Gan-dossi doesn't hold with that. He believes the alcohol extracts the scent of lemons almost at once and so he leaves it for no more than twelve hours. Then he makes a syrup by boiling up 1.6 litres of water and adding 1.2 kilos of sugar to it, stirring continuously until the sugar melts completely. He allows the syrup to cool to room temperature before tipping in the lemon peel and alcohol. Finally, he strains the mixture to remove the peel.

Eventually, we abandoned the vertiginous view for Gandossi's lovely kitchen overlooking the lake. Here he produced a bottle of the muted, ochre-coloured limoncello that he makes to his own recipe.

Out came the tiny glasses, filled to the brim with limoncello made only the previous day. This wasn't the sweet syrup served in tourist restaurants, it was a grown-up's drink, strong on alcohol but stronger still on the deep, exotic, powerful taste of lemons. What better glass to raise to the lemon and its extraordinary economic success from top to bottom of the Italian peninsula?

~

Limone's elaborate lemon houses were the magnificent expression of a long architectural tradition devoted to citrus growing. Ornamental citrus grown in northern and central Italy generally needs protection from frost, cold winds and heavy rain during winter, and pot-grown trees are often taken into the shelter of a purpose-built *limonaia* or lemon house. The history of these structures can be traced back to the sixteenth century, when brick sheds began to be built as permanent winter shelters for citrus. In *L'Idea della architettura universale* (1615), Venetian architect Vincenzo Scamozzi said these buildings must be over twelve feet wide and no more than sixteen feet long. He advised arranging the lemons and oranges according to their needs, remarking that oranges were satisfied with less heat and lemons required more sun. In 1618, the first purpose-built room was made for overwintering the Medici citrus collection in the Boboli Gardens in Florence. However, it was more of a storeroom than a *limonaia*, and often cluttered with statues that needed to be taken under cover as well as pots of citrus.

It has always been traditional to keep citrus under cover '*da Santa Caterina a Santa Caterina*', or from one St Catherine's Day to the next. This means that the trees still come in from the garden on 25 November, which is the saint's day of St Catherine of Alexandria, and go out again on St Catherine of Siena's day, 30 April. There are one or two exceptions to this rule, places like Buggiano Castello, a beautiful walled hamlet set among olive groves on the side of a hill near Montecatini Terme in Tuscany, where a freakish combination of conditions has made it possible to grow citrus outside throughout the year. Monks were the first to discover that they could espalier lemon trees against the walls of Buggiano's gardens, fields and houses in the fifteenth century, and they sold the fruit at local markets. Today all the inhabitants of Buggiano seem to share a passion for citrus cultivation, almost as if they had been interviewed before they were permitted to buy a house or take out the lease on an apartment in the village. Everyone seems to have enough land to make a garden filled with a combination of roses, brightly coloured flowers, vegetables and ornamental citrus trees, some of them grown as espaliers against walls, and some free-standing.

I went to Buggiano a few years ago to meet Luciano Disperati and Vanna Tintori, who have made their garden on two terraces overlooking the view. High walls enclosed it on three sides, trapping the perfume from the flowers of lemons, mandarins and a tree made up of two parts grafted on to the same rootstock, so that it bore both lemons and sweet oranges. A lemon grew against the wall of their house and Luciano told me he could lean out of the bedroom window and pick its fruit. Baskets of orchids hung among the branches of a mandarin, another witness to the exceptionally mild climate that Luciano described as 'a bit of a mystery.' He suggested that the marsh on level ground below the village attracted damp away from Buggiano, so they rarely experienced the fogs, frosts or cold Tramontana winds that afflicted villages on the same hillside only fifty metres below.

A *limonaia* is often an impressive building. Take the lovely rococo Stanzone degli Agrumi in the Boboli Gardens in Florence, which was built for Pietro Leopoldo in 1777 and is large enough to overwinter 500 pots of citrus. The building stands on the site of the menagerie used by the Medici to keep exotic animals given to them as diplomatic gifts. The giraffe and hippopotamus, the elephant and dromedary were long gone, and an elegant, 100-metre-long *limonaia* designed by Zanobi del Rosso took the place of their old quarters.

An even grander *limonaia* was erected in 1848 in the park of Castello di Racconigi, a vast castle built for the Savoy dynasty in Piedmont, up against the Alps in the far north of the country. They call the neogothic building designed by Carlo Sada a *serra* or glasshouse at Racconigi, and it forms part of the Margaria, a model farm used for agricultural research in the nineteenth century.

The Racconigi *serra* was famous all over Europe, not so much for the citrus collection overwintered there as for the cutting-edge design of its curved metal structure, inspired by Joseph Paxton's Great Conservatory at Chatsworth in England, and the vast collection of exotic plants housed on its lower floor. When I was there last it was the end of April and broody storks were sitting on enormous messy nests at either end of the roof. As I climbed the steps of the *serra*, I could already see that it had been emptied of pots for the summer. There was only one tree left inside the building, its modest frame dwarfed by a space over seventy metres long and twelve metres high, and yet when I pushed open the door I was enveloped in a warm blanket of perfume, the *zagara* on that small tree filling the entire building with its scent.

At the other end of the spectrum are simple agricultural buildings with little ornamental value. Whatever their status, they tend to be built with a solid wall to the north, south-facing windows, heavy wooden shutters and an earth floor that will absorb excess moisture from the air, preventing mildew. In the past fires were lit to warm the air on cold nights, but these days most *limonaie* have a modern heating system. When the temperature dropped to minus twenty-six degrees Celsius during the exceptionally harsh winter of 2011–12, modern underfloor heating in the *serra* at Racconigi saved the lives of all but two trees. Ornamental citrus trees used to be defined by the manpower needed to heave each enormous pot on to the sturdy wooden cart used to move it from the garden into the *limonaia* each autumn, a six-man tree being an average size. Smaller pots could be lifted with wooden poles inserted through the handles on each side, and pots too large to lift were moved on rollers.

Shifting the pots is a much easier job these days. At Villa Castello in Florence you will find a small tractor moving continuously between the garden and the *limonaia* in early May. The enormous double doors of the building will be open, and beyond them the L-shaped space is as big as the nave of a parish church. More than 500 trees overwinter there, one or two of them so big that their crowns brush the ceiling. There will already be a few dusty, leaf-littered spaces on the bare earth floor, but the transition takes weeks to complete.

Each tree in any citrus collection is labelled with a number that corresponds to an inventory number, to the number of the plinth it stood on in the garden and to its position in the *limonaia*. I've often wandered through the sun-drenched, dusty spaces of a *limonaia* in summer and seen a sheet of paper nailed to the wall or weighted down with a brick. On it will be a sketch, sometimes a very rough one, of the position of each pot when it is brought indoors in the autumn. The trees become accustomed to the temperature and fall of light in these places, and by replacing them in exactly the same spot each year, the gardeners minimize the trauma of the transition from inside to out.

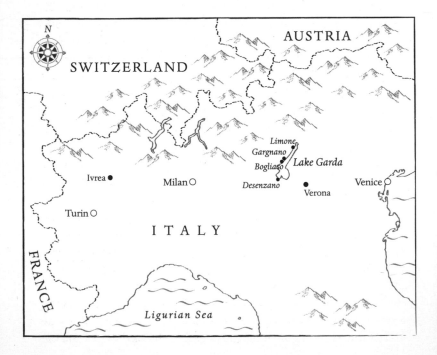

Battling with Oranges in Ivrea

Ivrea may be in Piedmont, Italy's far north, but its name is inextricably associated with oranges. It's not that the citizens of Ivrea grow them, or even sell them. Quite the reverse. They throw them away. Ever since the nineteenth century oranges have been used in vast quantities as ammunition during the violent, hilarious and unbelievably sticky Battle of the Oranges that marks the end of the carnival season.

The battle was an early-nineteenth-century invention. Until then, Ivrea's carnival was much like any other. With nothing to look forward to but the restrictions and privations of Lent, people took full advantage of *carnevale* to enjoy their last few days of freedom. On the Saturday, Sunday and Monday before Shrove Tuesday the world turned upside down. Masters were obliged to take orders from their servants, men were permitted to dress as women and women put on masks and did whatever they fancied. The party was continuous and many of the revellers dressed up as priests, lawyers or other figures of authority, and then behaved in shocking and humiliating ways. The prospect of forty days of fasting brought food to the forefront of everybody's minds, making gluttony and copious drink a habitual part of the festivities. And under the cover of carnival, old feuds were settled, so that fighting, or watching other people's fights, became another element of the entertainment.

In the midst of all this raucous revelry, Ivrea's aristocracy had always persisted in imitating a genteel ceremony practised in other more sophisticated European cities such as Nice. This found them standing on the balconies of their *palazzi*, high above the street, and – in a great show of beneficence – throwing flowers, sugared almonds and other tiny gifts to the crowds below. In the eighteenth century some of Ivrea's most fashionable families began to spend part of the winter in the mild climate of the Italian Riviera and the

Côte d'Azur. They came home for the carnival and sometimes they brought oranges with them. Oranges were thought rather chic in eighteenth-century Piedmont and it was natural to add them to the trifles they threw to the carnival crowd. However, in a gesture that perfectly encapsulated the inverted spirit of carnival, the people in the crowd threw the oranges straight back. This exchange began as a game, but it became increasingly energetic, then aggressive, and eventually it degenerated into a fight so wild and uncontrolled that the authorities felt obliged to step in. By 1850 orange throwing was illegal; of course, this made it even more appealing and the Battle of the Oranges was soon established as a traditional part of Ivrea's carnival.

I knew no one in Ivrea and so when I first began to think of going to the battle I sent a message to the tourist board. I received an answer at once. At first the message appeared entirely formal. 'Here's the website for hotels,' it said, and 'we hope you enjoy your stay.' At the bottom of the message was something else, something written in a quite different tone. 'That said,' it began, 'your message intrigues me . . .' By the end of the message I'd been invited to stay in the middle of Ivrea for the duration of the carnival. And that's how I came to meet Lucia Rossi for the first time on the railway platform. 'First things first,' she said, 'you'd better put this on.' 'This' was a red, stocking-shaped hat of the kind that an elf might wear. Lucia already had hers on and she explained that with a hat on my head I'd be seen as a supporter of the revolution and so no one would throw oranges at me. What revolution? I thought, cramming the thing self-consciously on to the back of my head. It fell off at once, but as we walked through the quiet, sunlit streets, it was hard to believe that I was in any danger.

Lucia was the perfect guide and she took the opportunity to explain a few of the fundamentals of the Battle of the Oranges. She told me the battle and the rituals surrounding it were an amalgamation of two important events in Ivrea's history. The first was the medieval legend of the miller's daughter, La Mugnaia. She is said to have rebelled against Baron Raineri di Biandrate, the feudal overlord of Ivrea and a great believer in the *ius primae noctis* that gave him the right to take the town's brides into his own bed on their wedding nights. The miller's daughter defied the baron, chopped off his head and displayed it triumphantly from the battlements of his castle. These events may be legendary, but it's true that the citizens of Ivrea rebelled against the baron in 1194, stormed his castle and banished him for ever, and it's this revolutionary spirit that still informs their carnival. The second event commemorated is the Napoleonic occupation of the town at the beginning of the nineteenth century. Lucia was about to explain all this. but as we climbed the hill and turned the corner to the main street we were met by the lens of an enormous camera. 'Oh dear,' Lucia said, 'now your picture will be published next year in the catalogue of traitors.' Blimey, that's a bit serious, I thought, I've hardly arrived. And so, too late, I began to understand the symbolism of the red hat bundled up in my pocket. It is modelled on the Phrygian cap worn during the French Revolution. Wear it and you're on the side of the revolutionary angels. Without it, you're one of the tyrants and you deserve to be bombarded with oranges. I put it on at once, but I was too late, and the following year when Lucia sent me the programme for the carnival I found my unsuspecting face among those of other traitors.

Dramatic scenes from the Battle of the Oranges are easy enough to find on the internet and I arrived expecting to find a festival focused entirely on citrus warfare. Before long I realized that the battle is just one element in an intricate production beginning on 6 January and ending on Ash Wednesday, and built around so many different plots, timelines and ceremonies that you'd need to be an anthropologist to interpret them all. The battle is the most famous of the ceremonies associated with the carnival. It attracts the younger generation, draws thousands of tourists to the city each year and, according to Lucia, threatens to dominate the entire proceedings. But the carnival committee fights constantly to keep all the different strands of the event alive. And things were certainly lively in Piazza Maretta, where the Fagiolata di Castellazzo ('Castellazzo's bellyful of beans') was well under way. We pushed to the front of the crowd to find a temporary roof sheltering three cauldrons so enormous that the chefs had to climb ladders to stir them. Oversized saucepans, tiny chefs, lots of steam and crowds of helpers wearing little red hats. Had we blundered into a fairy tale? No, there was nothing imaginary about the steaming bowl of beans in my hand. 'Smells good,' I said to Lucia, although it wasn't yet ten in the morning. She explained that the beans had been boiled up all night with the 'fattiest parts' of pigs. 'There are trotters in there, lots of skin and sausages,' she said brightly. This meaty dish of beans would have tasted even better when *carnevale* really was a long farewell to meat, and insecure food supplies and the dictates of the Catholic Church made eating an anxious cycle of fasting and feasting.

There was still no sign of battle, but as we walked through the city squares I noticed boxes of oranges piled high at regular intervals along their edges. The piazzas are Ivrea's battlefields and Lucia explained that hundreds of *arancieri*, or orange throwers, would gather there to confront the enormous horse-drawn floats carrying teams of warriors representing the tyrant's armed guards. It costs each of the warriors about €500 to secure their place, and they must also buy helmets, protective clothing and their share of the four tons of oranges that would be their ammunition.

All the *arancieri* belong to different teams that correspond to the various districts, or *rioni*, of Ivrea, and each *rione* has command of a piazza, or part of one. Anyone can be an *aranciere* as long as they pay a contribution to the relevant *rione* for the cost of the oranges. And if visitors want to join the battle at the last minute they're welcome, as long as they only throw oranges picked up off the ground. Somewhere I'll have to admit, and it may as well be here, that I proved too much of a coward to join in. But if you'd seen the speed of the fruit as it hurtled through the air, and heard the wet thwack as it hit someone in the mouth, the eye or the ear, you'd be cowardly too.

Oranges come to Ivrea from all over southern Italy and Sicily. They are the sweepings of the packing houses and yet there are strict rules about the oranges that can be used in the battle. They mustn't be too big, because that would make them heavy and awkward to handle, and they mustn't be too small, because that would be fiddly. And all of the fruit must be washed and packed into boxes small and light enough to hand up on to a float. The packing house I visited in Sicily sends spoilt fruit to Ivrea every year.

We returned to Lucia's apartment for lunch, and while we were there the first floats began to move along the street below. Standing on her tiny first-floor balcony, we were perfectly placed to oversee events. Lucia lives at the bottom of a short, sharp hill. It's here that a special police force, the 'zoological police,' gather to check the horses' well-being and ensure that no more than seven passengers ride on each float as it goes up the hill. The zoo police inspect the horses thoroughly, running hands over their sleek bodies and then looking up at the float to count heads. They may have seen only seven, but from our vantage point on the balcony we could see the other five members of each team crouching down inside the float.

The first float entered the main piazza at 2.30. There was a roar from the *arancieri* and the battle began at once. Some of the warriors on the floats wore heavy modern helmets with perspex visors or metal grilles over their faces. Others had traditional leather helmets that had been tailor-made for them by local craftsmen. Each float was designed around a theme. The most striking, I thought, had a gallows at the front and was manned by a team of hangmen. There was something monstrous about all of the warriors on the floats, with their oversized padded shoulders, their huge, helmeted heads, invisible faces and the vicious overarm action they used to throw oranges with brutal force at the bare heads of the *arancieri* on the ground, who wore no protective clothing whatsoever. They were equipped with only the cloth bag they filled with oranges. They gathered at the base of the float and threw them at short range, straight at the helmeted heads on board, or at any area of exposed flesh, particularly on the upper arm, where a blow really hurts, I was told. The oranges rained down on warriors and *arancieri*, bursting open with the force of impact and spraying juice that soaked into clothes and coated every surface. I've been told that somebody once experimented by putting their oranges into a freezer before the battle, effectively turning them into lethal weapons, but that's illegal now.

Nevertheless, the faces of the *arancieri* were soon running with blood as well as orange juice, and inside their helmets the tyrant's guards were both blinded by juice and part suffocated, but the fight went on, accompanied by screaming, bellowing, battle cries, taunts and chanting. And all the while the horses under fire were restless but utterly calm. Nobody ever threw oranges directly at them, although they occasionally took a blow from an orange that had ricocheted off the float.

The battle stopped suddenly and the float moved off towards the street. The *arancieri* relaxed behind the protective netting that surrounded the piazza, drank hot *vin brulé*, Piedmont's version of mulled wine, nursed bleeding fingers and sang rousing songs. There were thirty-eight floats in all and so it wasn't long before another one arrived. By now the ground was covered in a thin layer of squashed oranges. As the day wore on I watched this turn into a thick slurry of orange pulp and horse manure. No one wants this muck inside their shop and so most of Ivrea's streets are shuttered during carnival. Just a few brave bars and food shops remained open, their floors protected by multiple layers of cardboard and their walls by plastic sheeting. By dusk you walked at your peril in the piazzas and the streets immediately surrounding them, each footstep a squelching, sucking challenge. A little further away from the piazza, the cobbles were glazed in a transparent and highly polished varnish of juice. It was so slippery underfoot that even *arancieri* in heavy boots moved like skaters across them. And wherever you went, you couldn't escape the *odore del carnevale*, a unique combination of oranges and horse dung. It's unmistakable and, according to Lucia, simple washing won't take it out of your clothes. You have to wash everything two or three times by hand before you even put it into the washing machine. There's money to be made here and this year, for the first time ever, a launderette was offering an overnight washing service. Outside, they had rigged up a makeshift cubicle where customers could strip off their dirty clothes.

The *arancieri* had an unlimited supply of oranges in boxes on the piazza, but the floats had to replenish their stock in the streets between battles. This was a moment of relaxation when the guards pulled off their helmets, revealing flushed faces that weren't monstrous after all, and the horses stood briefly at ease, rugs flung over their backs. Someone handed boxes of oranges up to the float, where they were tipped into easily accessible gullies built along its edge. Someone else swept the floor clean of squashed fruit and others took the chance to rinse their helmets out with water and adjust the bandages protecting their wrists. When everything was tidy, the ladder was pulled up at the back of the float and it set off at a spanking pace.

The floats move in a circuit between the different piazzas, and so does the carnival procession led by La Mugnaia in her gilded carriage. The figure of the Mugnaia has been a crucial element of Ivrea's carnival ever since 1858 and there are people in the town who could tell you the names of every Mugnaia there has ever been. There is no official process for choosing the Mugnaia, but being invited is the greatest honour imaginable for a citizen of Ivrea. You can only do it once but the recognition lasts for the rest of your life. It's not something you can apply for, and in fact that would be counterproductive. Anyone can be Mugnaia, just as long as she has access to enough money to pay for all the parties, gifts, dresses and other expenses that the privilege entails. According to one ex-Mugnaia, the carnival season cost her about as much as the wedding of her daughter – and in Italy that's saying something. The identity of the Mugnaia is absolutely secret until the Saturday of the carnival. Only her close family are allowed to know who she is, and even the dressmaker in charge of her costume isn't allowed to see the client's face.

The Mugnaia is accompanied everywhere by Il Generale, a general riding a handsome horse who is said to have been appointed by the Napoleonic administration to keep order in Ivrea during carnival. The entourage that follows them in the procession is composed of an extraordinary combination of figures from different historical periods. There are soldiers on horseback in Napoleonic uniform, children on ponies in Renaissance costume, musicians playing fifes and drums, standard bearers and many others following behind. All this might sound like a sterile historical pageant, but if you had stood among the dense crowds that lined the streets of Ivrea as the procession passed, listened to the applause and the laughter, the music and the passionate shouts of '*Viva La Mugnaia!*', you would understand that the spirit of Ivrea's carnival was alive and well. Whenever the procession reached a piazza, snow ploughs cleared a path through the slush of horse dung and squashed fruit for the Mugnaia's carriage. The contrast between her immaculate white dress and the sticky, stained, injured *arancieri* that made way for it became ever more striking as the battle progressed.

When darkness fell, the battle ended and I went back to Lucia's lovely apartment. Fifes and drums played their way up the street outside and behind them came the snow plough that scraped the piazzas and surrounding streets clean each night. Bruno, Lucia's partner, came home. He was an *aranciere*, and it showed. He had one black eye and a big bruise on the opposite cheek. On the second evening, Bruno called to Lucia from outside the front door. 'I'm a bit bruised,' he yelled, 'so don't overreact.' And in he came, with two black eyes this time. On the third evening of carnival Bruno's injuries appeared much as they had the day before, although when I asked how he felt he said, 'Even my hair hurts.' Never mind, carnival had finished, and we stood in darkness on the balcony to watch its funeral procession pass in the street below. All the riders had dismounted and hundreds of people walked slowly behind them, their eyes cast down. Nobody spoke; the only sound was of hoof beats and the sad shuffle of feet.

And when it was all over, 400 metric tons of orange slush had to be taken to the municipal composting unit. And another fifty metric tons of fruit boxes had to be gathered up and disposed of. The clear-up is known as the *emergenza arancia*, 'the orange emergency', in the garbage trade, and it involves gangs with brooms and pressure hoses, fifteen snow ploughs and five tractors fitted with brushes for scrubbing the streets. And far below, at the bottom of the hill, oranges drifted downstream like bright bubbles in the ice-green water of the River Doria.

Green Gold

Calabria and the most valuable citrus in the world

⟶⟜◌⟞⟵

Wherever citrus trees are gathered together, whether in open ground or the shelter of a *limonaia*, they cross-pollinate and over time varieties develop that are peculiar to their setting. The unusual qualities of a distinctive fruit such as *Chinotto di Savona* made it popular for a time, but not popular enough to resist changing fashions. In Calabria, however, there are two kinds of citrus fruit with long commercial histories and unique qualities that protect them and ensure that they are always highly valued.

Calabria is Italy's deep south, its southernmost tip, quite distinct from the rest of the country. Drive there from Rome and the heat accumulates in the car as you travel, like an extra passenger picked up somewhere south of Naples. If you fly to Calabria from Britain the complex journey generally has you changing planes in Milan or Rome. 'Or I could fly into Naples and take a train,' I once said to an old friend in Rome. 'No,' she replied, in that special, disgusted voice she reserves for cities other than her own. 'You don't want to do that. You'd have to catch a cab to the station and you'd just get ripped off.' I've spent many happy times in Naples over the years, but on that occasion I took myself to the antiseptic safety of Milan airport, where icy winter rain fell on the sour tarmac. Businessmen turned up the collars on their macs and hurried towards the exit and the offices and factories beyond it, but I had time to kill before the flight to Lamezia and I wandered over to the Motta bar. They have a machine in there for making *spremuta d'arancia*, freshly squeezed orange juice, and the whole place smelt of oranges, a good omen for a traveller on a citrus pilgrimage.

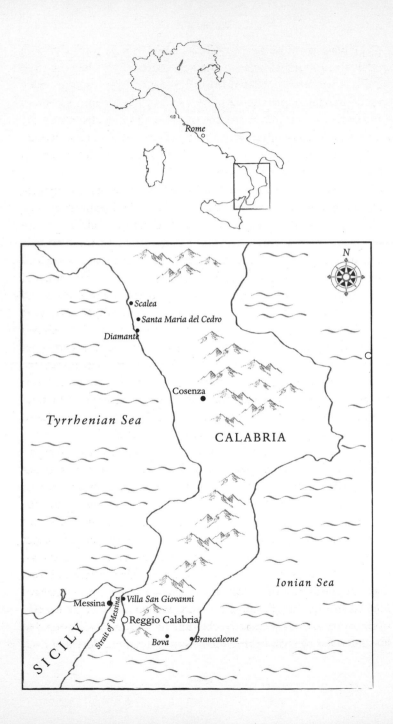

The plane from Milan to Calabria turned its nose towards Africa and flew in a straight line down the west coast of Italy. I had expected sun, but we bounced down the length of the peninsula in thick cloud, arriving somewhere above Lamezia airport in a storm so intense that it was impossible to land. The plane circled above the airport and I caught glimpses of a landscape of sodden hills through ribbons of racing fog. Eventually we landed in a tepid deluge, much warmer than the winter rain in Milan.

The first of the region's unique and valuable fruits is bergamot (*Citrus bergamia*), the product of a natural cross-pollination between a lemon tree and a sour orange that occurred in Calabria in the mid-seventeenth century. Essential oil can be extracted from the bergamot's fruit, and although its extremely high value has inspired many attempts to grow it elsewhere, bergamot is like an animal in its chosen territory: it thrives and fruits successfully only on a thin strip of coastline that runs for seventy-five kilometres from Villa San Giovanni on the Tyrrhenian coast to Brancaleone on the shores of the Ionian Sea. Here the tree grows tall and strong, and bears such heavy crops that its brittle branches often snap under the weight of oily fruit. Take it away from its home ground and you make it a perpetual invalid, incapable of tolerating the cold or weathering strong winds. Any essential oil extracted from fruit produced outside Calabria's bergamot belt is of inferior quality.

Calabria is bounded on its east side by the Ionian Sea and on its west by the Tyrrhenian Sea. Its southernmost tip almost touches Sicily across the Strait of Messina and the Apennines stride down its centre, dividing east from west. Mountains have always cut the region off from itself. You only have to read Edward Lear's account of walking through Calabria in the mid-nineteenth century to get an impression of the difficulty of the terrain. Lear is best known today for his nonsense poems, and yet he was also an intrepid traveller and the author of vividly descriptive illustrated travel journals. He had already lived in Rome for ten years and had been on expeditions to Abruzzi and Molise by the time he set out for Calabria. He had intended to spend three months in the region, devoting each month to one of its three provinces. However, political unrest in 1847 and a general uprising against Bourbon rule in 1848 obliged him to cut his stay short.

Lear set out to visit places in Calabria that had been ignored by earlier English travellers. He walked all over the province of Reggio Calabria in the south of the region accompanied only by an English companion, a local guide and a horse to carry their luggage. Lear must have encountered a number of practical difficulties during his journey. The first challenge would have been the extreme gradients of the landscape. In winter ferocious torrents descend from the mountains, cutting villages off from one another. In summer the deep channels that they gouge from the land would have been a formidable obstacle. Lear describes *fiumare*, as these dry torrent beds are still called, as 'Lonely places of devastation . . . blinding in their white or sandy brilliancy, barring all from view without their high cliff-sides, and recalling by the bare tract of ground right and left of their course how dismal and terrible the rage of their wintry watery occupant has once been throughout its destroying career'.[1]

Lear's second problem would have been linguistic. He spoke some Italian but that would have helped him very little in the province of the regional capital, Reggio Calabria, where they speak a dialect derived from ancient Greek. The shores of southern Calabria and eastern Sicily once belonged to Magna Graecia and they remained part of the Greek Empire until the area was annexed by the Romans in the third century BC. Although the Romans brought Latin to the region, they allowed the Greek inhabitants to continue speaking their own language. *Grecanico*, an archaic dialect derived from medieval Greek, is still spoken by approximately 5,000 people in Reggio Calabria and the province surrounding it. Many of them live in the remote mountain villages that Lear visited. I'm told that the little village of Gallicianò, high above the Ionian Sea, is still inhabited almost entirely by *Grecanico* speakers. And, incidentally, 'ndrangheta, the name of the local Mafia, derives from the *Grecanico* for 'heroism' or 'manliness'.[2]

~

The idea of Calabria as difficult terrain for travellers still has its advocates and when I announced my plan to go there some of my friends in central Italy tried to dissuade me from hiring a car. They told me the roads were uniquely dangerous and made dark allusions to the 'ndrangheta. 'Trust no one,' they said finally, as if I were going to a distant continent, and for a moment I was nervous. And it's true that Calabria has long been considered the most isolated and impoverished part of Italy. The region is in the grip of the 'ndrangheta, and I've spent enough time there to hear a great deal about political intrigues and crime. The dominance of this corrupt political climate over people's lives was brought home to me by a citron producer in northern Calabria. He recounted his fascinating, shocking and immensely detailed story as he drove me along a bumpy track through his citron grove, and it unfolded against the sound of thrashing wipers and rain hammering on the roof of his elderly Fiat Cinquecento. I can't repeat anything he told me because after a while he said, 'You won't write about this stuff, will you?' 'No, of course not,' I said. Suddenly he jammed his foot on the brake and we slewed to a halt. He cut the engine and in the silence the only sound was of drumming rain. Then he turned to face me and said, 'I'm eighty-one years old. You don't lie to an eighty-one-year-old. Will you write about it or not?' 'No,' I replied. And I won't.

Calabria's physical problems are as dramatic as its political ones. It has been devastated again and again by lethal earthquakes and afflicted by both drought and the flash floods that follow torrential rain. In winter the rain often causes dramatic landslides, as in 2010, when the inhabitants of the small town of Maierato were forced to evacuate their homes as mud destroyed houses, roads, power lines and anything else in its path.

This harsh and beautiful landscape, beset by extreme weather conditions and overshadowed by political corruption, is home to the most valuable citrus fruit in the world. Simon Barbe, a celebrated parfumier in seventeenth-century Paris, made the wild suggestion in his book *Le Parfumeur Royal* that the tree was first created by grafting a lemon on to a bergamot pear. Another legend links the bergamot's name to the Catalonian city of Berga in Spain, and suggests that Christopher Columbus brought it to Europe in the fifteenth century from the Canary Islands. Only one thing is certain: its first appearance anywhere in the world was in the mid-seventeenth century in Calabria.

Drive south from Reggio Calabria towards Bova Marina and you can see bergamot trees on the narrow plain between the foothills of the Aspromonte mountains and the sea. They grow in glistening, dark green swathes between dramatic plugs of volcanic rock and on narrow terraces cut from a sheer cliff face. The trees have large glossy leaves similar to a lemon's and bitter fruit that ripens from green to yellow and is the size and shape of an orange. Anything goes in a bergamot grove. Trees are pruned very lightly only once a year and some of them grow to over four metres high. They are carefree, liberated, untidy and entirely organic, the hippies of the citrus world. It is the essential oil stored in the pores just beneath the surface of the skin that makes bergamot so valuable. Ever since the beginning of the eighteenth century the principal and most lucrative use of this oil has been as a fixing agent in the perfume industry. The addition of bergamot oil makes a perfume last longer and brings all its other elements into harmony, rather like the conductor of an orchestra.

Castagnaro, Femminello and Fantastico are the varieties of bergamot cultivated here. Castagnaro is the oldest of the three. It's a big tree and it doesn't produce every year, but when it does there is a bumper crop. Fantastico produces an abundance of essential oil, although the oil from the Femminello is of the highest quality. In bergamot groves close to the sea the trees' white blossom begins to open at the end of March. The *zagara* season extends into April further inland, where the groves are not quite so warm and sunny, and the harvest begins in October on the coast and lasts until December inland.

When bergamot first appeared in Calabria it was immediately appreciated for its blossom, which has a stronger scent than any

other *zagara*. The bitter fruit was not considered edible, but bergamots were planted as ornamental trees in the gardens of villas in its
homeland near the regional capital, Reggio Calabria. Visitors appreciated the extraordinarily powerful perfume of the fruit's essential
oil, and when the English writer Henry Swinburne travelled through
Calabria in the late eighteenth century he remarked that the inhabitants of Reggio 'carry on a lucrative traffick with the French and
Genoese in essence of citron, orange and bergamot', which they
'squeezed into a phial and sold for fifteen carlines an ounce'.[3] It's
likely that these Genoese and French customers were associated
with the perfume industry in Grasse.

~

The history of bergamot is intimately connected to the life of Giovanni Maria Farina, born in Piedmont in northern Italy in 1685. For
generations the Farina family had been masters of the art of distilling the alcohol that is used to carry the fragrance of a perfume.
Many people distilled spirits to drink, but very few understood the
art of distilling pure alcohol, or ethanol. The complex technique
was first discovered by the Arabs in North Africa, although the
knowledge was soon suppressed because alcohol was prohibited.
However, the skill of distillation continued to be known in southern
Italy, and came via Sicily to Piedmont, where it lived on in the workshops of the Farina family. Giovanni Maria Farina's grandmother
came from a long tradition of perfume making and by the time
Giovanni Maria was fourteen she had already noticed that 'He
divides people into those who smell good and those who smell bad.
Life will not be easy for him, but his nose is as quick as his mind.'[4]
He proved exceptionally skilled at the art of distillation and of combining alcohol with fragrances extracted from fruit and blossom. As
a young man Farina emigrated to Germany to pursue his profes-
sion. In 1708, when he was only twenty-three years old, he invented
a bergamot-based perfume. He named it Eau de Cologne in honour
of the city where he lived.

In a letter to his brother, Giovanni Maria described his new invention as 'the scent of a spring morning in Italy, of mountain narcissus and citrus blossom after rain'.[5] The top note of the perfume, borne by evaporating alcohol, enveloped the wearer in this aura of the Mediterranean. For Giovanni Maria, this was the smell of home; for his northern European customers the perfume conveyed the exotic essence of everything they yearned for in the sunny south. Its middle and base notes combined with the wearer's skin to produce an individual scent for every person who wore it. Eau de Cologne was a huge commercial success and an international bestseller. It became the perfume of the great houses and royal courts of eighteenth-century Europe, and it might have been the first thing you noticed when meeting King Louis XV, Napoleon Bonaparte or Mozart. Voltaire described it as 'a fragrance that inspires the spirit', and if you had visited Goethe, that great lover of southern Italy and its citrus trees, you would have found him writing beside a box of rags soaked in the stuff. Soldiers fighting in Germany during the Seven Years War spread Eau de Cologne's fame by taking bottles of it home with them to Austria, France and Russia. When perfumes were banned in France during the Revolution, Eau de Cologne survived by temporarily rebranding itself as a medicinal tonic or elixir. It went on to dominate the market for 300 years, leaving little opportunity for the success of any other perfumes.

Farina was determined that every batch of his Eau de Cologne should smell exactly the same. This was almost inconceivable in the early eighteenth century, when it was generally accepted that perfumes would change in response to variations in their natural ingredients. He overcame this problem by composing each perfume from a combination of essences extracted over a broad time span, adjusting the mixture laboriously until he achieved a uniform effect every time. Bergamot oil was the essential ingredient in Farina's formula and he could buy it only in Calabria. At first he must have bought tiny amounts of oil from individual producers, because there is no record of bergamot being grown on a commercial scale until 1750, when Cavaliere Nicola Parisi planted the first bergamot grove at Giunchi, on the outskirts of Reggio Calabria.

By the time Edward Lear travelled through Calabria in 1847, 3,000 acres of the Calabrian coastline were covered in bergamot groves. He described Reggio as 'one vast garden' and saw the 'thick verdure' of oranges, lemons and bergamots that 'stretched from hill to shore as far as the eye could reach on either side, and only divided by the broad white lines of occasional torrent courses'.[6]

At first Farina bought oil that had been extracted in Reggio Calabria and sent to him in Germany in copper vessels. The quality of it varied because extraction is an extremely sensitive process, and the oil can be spoilt by the slightest change in the process, or even the draught through an open door. Farina noticed that bergamot oil was reacting with the copper vessels in which it was stored, and this also affected its quality. After that he insisted the oil should be stored only in glass. 'I will accept ceramics too,' he wrote to his dealer, 'but sealed only with linen and beeswax, not with resin.' Despite these stipulations he was still dissatisfied by the variation in the quality of essential oil extracted. He tried to tackle the problem head on by contacting the bergamot farmers and giving them minutely detailed instructions on watering the trees, 'as the whole scent depends on this'. However, there was little he could do at long distance to ensure they followed his instructions, and the quality of the oil continued to be unsatisfactory. Eventually he took what control he could by importing whole fruit from Calabria and extracting the oil from it himself. The arrival of the first consignments of fresh bergamot made headline news in Cologne, and women brought small children to the distillery, lifting them up to its high windows so that they could inhale the intense, health-giving scent of the fruit.

Other manufacturers tried to imitate the Farina family by mixing their own versions of Cologne, and in 1803 Wilhelm Mülhens, another parfumier in the city, attempted to break the family's monopoly by purchasing rights to the name of Farina from a dubious family member with no genuine link to Giovanni Maria. The Farina family took Mülhens to court and the ensuing battle endured for generations. Eventually Mülhens's grandson was forced to choose a new name for their company and for the cologne they produced, and in 1881 Mülhens's version was renamed 4711, after the street number assigned to his house in Cologne on the eve of the French Revolution, when the city was occupied by the French and had already been seized by revolutionary troops. Throughout the nineteenth century Mülhens's 4711 was Farina's chief rival, but it was not until the twentieth century that it eventually cornered the popular market and outsold the original Eau de Cologne.

Send an email to Farina & Co. today and answer will come from Johann Maria Farina, company director and direct descendant of Giovanni Maria Farina. He told me that exactly the same formula is still used, and they have even held on to the original premises. Johann Farina was kind enough to send me two small bottles of Farina Eau de Cologne. I found I could recognize the sunny, optimistic top note that bergamot oil brought to the perfume. A few drops on my wrist filled the room with bright scent.

~

Ever since the nineteenth century bergamot oil has also been used to give a distinctive aroma and flavour to one of the most popular teas in the world. Earl Grey is made from a blend of dark China teas treated with the bergamot oil or peel. It is always said to have been named after the second Earl Grey of Howick, Britain's prime minister from 1830 to 1834. The story revolves around a diplomatic mission to China, when Grey is supposed to have rescued the son of a Chinese mandarin from drowning. The mandarin was said to have been so grateful that he sent a box of bergamot-scented China tea to Grey in London. There are several things in this story that don't add up, not least that Grey never set foot in China and bergamot grew only in Calabria. Nevertheless, it is certain that the English developed a taste for scented teas at about this time, blending Indian and Sri Lankan leaves, dousing them with oil of bergamot and naming the mixture Earl Grey in honour of their prime minister. Earl

Grey tea is still produced in the same way and consumed with undiminished enthusiasm in England.

By the mid-nineteenth century bergamot was produced on an industrial scale, and harvesting the fruit and extracting its oil involved the entire community in the coastal villages between Villa San Giovanni and Brancaleone. The job of picking was given to boys and the youngest men. They used pruning knives to remove any stalks that might pierce or damage the skins of other fruit, and then placed it gently in a basket lined with sacking to give it further protection from bruising or chafing that could cause the oil to ooze from its skin and be wasted.

When a basket was full it weighed about thirty kilos. It was the women's job to hoist the baskets on to their heads and carry them in from the fields to the *fabbrica*. Here, the working day began at 4 a.m. and finished at 4 p.m. The skin of a bergamot is so full of oil that it will begin to seep out at the slightest pressure, so initially it was extracted simply by pressing and turning the peel of the fruit against a sponge. The sponge was then squeezed into a glass phial and the liquid was left to settle and separate. The oil produced by this ancient method is unrivalled, and according to one 'nose' in the perfume industry, the difference between it and oil extracted mechanically is 'just the same as the difference between a bull and a bullock'.

In 1844 the *macchina calabrese*, the 'Calabrian machine', was invented by someone called Nicola Barillà, and this revolutionized the laborious extraction process. A few bergamots of similar size were placed between two metal cups. The lower cup was covered in spikes to hold the fruit still and the upper one was armed with sharp blades. Two men took it in turns to operate the handle that rotated the cups, and the combination of pressure and the movement of the upper cup made oil and water spray out of the peel and fall into a copper bowl. Finally, the mixture of grated peel and oil would be strained through woollen sacks that were hung from a rack and left to drip into another copper bowl. All of the copper bowls used during the extraction process were lined with tin to prevent the oil reacting with the copper.

The *macchina calabrese* continued to evolve, becoming increasingly sophisticated over the years. Bergamot offered higher returns than any other crop in the country, though historically very little of this money was seen by the people who worked hard to cultivate the trees and pick and process the fruit. Some profits were made by landowners, many of them living in northern Italy, but in the nineteenth century the real money was earned by British companies based in Sicily.

Most of these companies were set up at the beginning of the nineteenth century, when thousands of British troops were sent to bolster the forces of Ferdinando I, the Bourbon king of Naples and the Two Sicilies, who fled to Sicily from Naples on the eve of Napoleon's invasion. Lord Nelson delivered Ferdinando and his entourage to Palermo on HMS *Vanguard*, and British soldiers were sent to the western side of the island. Their presence encouraged several English merchants to go to Sicily and get involved in the three main industries: sulphur mining, citrus fruit and marsala, a fortified wine much appreciated by Nelson.

One of the first foreigners to arrive was Benjamin Ingham, the youngest son of a family of textile merchants from Leeds, who came to Sicily in 1806. He staked his claim and built his business empire on these products before the involvement of criminal syndicates that would eventually evolve into the Mafia. Ingham employed five of his nephews to help him. His reputation for hard-nosed determination was enhanced when one of them died and it was rumoured that he wrote to their mother, saying, 'Your son is dead, send me another.'[7] He spoke Italian fluently with a mixture of a Sicilian and a Yorkshire accent, and profits reaped from marsala, citrus and other traditional products made him one of the richest and most powerful men in Sicily. He was part of the business consortium that brought the first steamship from Glasgow to Palermo, thus transforming the citrus trade between Sicily and the USA. His most lucrative markets were in America, and a nephew was sent to oversee the arrival and distribution of Ingham's goods. This gave the company a considerable advantage over other Sicilian citrus traders.

Much of the profit from bergamot oil was reaped by William Sanderson and Sons, a British company based in the Sicilian port of Messina. Sanderson was a sailor serving under Nelson during the Napoleonic Wars. Wounded in battle, he was taken to Messina to convalesce. While recovering from his wounds, he watched citrus fruit being brought across the Strait from Calabria to be shipped out of Messina and came up with a business idea to keep him in that beautiful place. He must have observed that bad weather sometimes delayed the crossing between Sicily and the mainland, and seen that this could result in damage to fruit, or even its loss. He would also have known that the essential oil was extracted from bergamot as soon as it was picked and realized that it suffered no harm if its journey to Messina was postponed.

Sanderson cleverly established a company trading in the essential oils extracted from citrus fruit of all kinds, but by far the most valuable of them was Calabrian bergamot oil. He founded William Sanderson in 1817, and was soon exporting essential oils to Britain and its ever-expanding empire all over the world. Over the years Sanderson dealt in a variety of citrus derivatives, including pectin, lemon juice concentrate and essential oils. The company remained in the family until William's nephew sold it to Italian investors in 1906. Two years later the original factory buildings were destroyed in an earthquake, but Sandersons continued to trade from new premises under the same name until 1981.

~

Calabria's bergamot belt is just across the water from William Sanderson's home in Messina, and he would have been a familiar figure in towns such as Villa San Giovanni at its western end. To a stranger, the place seems to exist today only for its port. Not much happens there until a ferry from Sicily docks, and then the streets are suddenly gridlocked with juggernauts impatient to reach the open road and begin the long haul north. The promise of a meeting with Ezio Pizzi, bergamot's ambassador, brought me there for the first time. As I drove into town I noticed an enormous hoarding advertising *Degustazione dei prodotti tipici e al bergamotto*, an entire building devoted to tasting local food and products flavoured with bergamot.

Pizzi took me out of town again almost as soon as I arrived that night, and as we sped down the dark motorway towards Reggio Calabria in his car, he urged me to think about the extraordinary fact that bergamot oil combines 368 separate chemical components. Its traditional role in the perfume and tea industries is ongoing, but bergamot oil's antiseptic and antibacterial qualities mean that it is also used in hospitals all over the world as a disinfectant to combat superbugs and post-operative infection, and in dentistry, ophthalmic surgery, gynaecology and dermatology. These qualities were first discovered by a local doctor called Francesco Calabrò in the early nineteenth century. He was often called on to treat the workers employed to extract essential oil from the skins of bergamot. In those days the fruit was peeled with a razor-sharp, broad-bladed knife, and despite protection the workers often cut themselves. Yet Calabrò was fascinated to observe that these cuts never became septic and that they healed extraordinarily fast. From this empirical evidence Calabrò deduced that the oil itself had antiseptic and healing properties.[8] Those properties were further built upon by Ferdinando Bergamo. An army doctor in Reggio Calabria, he experimented with the oil as both a preventive and a cure for scabies. In April 1853 he was called to treat a family with eight children and a young male servant, all of them suffering from the intense itching and skin rash that are the main symptoms of infestation by the microscopic mite that causes the condition. Inspired by the work of Francesco Calabrò, he decided to prescribe treatment with bergamot oil. When he returned eight days later the entire family was cured.

This success gave Bergamo the idea of using bergamot to treat the scabies that was endemic among soldiers billeted nearby. His approach consisted simply of rubbing a mixture of bergamot and almond oils all over the body at regular intervals. Diluting the bergamot with almond oil prevented it from irritating the tender skin of the younger soldiers. Bergamo's system replaced an ineffective and very much more complex treatment involving a sulphur bath, which did nothing for the quality of the air in the barracks or the condition of the soldiers' skin.[9]

Research into the use of bergamot as an antiviral treatment for AIDS and as an antidepressant is ongoing, and the oil is used in the manufacture of creams for treating wounds and cuts, psoriasis, chickenpox, oily skin, dandruff, eczema, acne and cold sores. It is also widely used as a perfume for soaps, deodorants, toothpastes and shower gels.

By this time I was dazzled by bergamot's potential, and Pizzi was already moving on to explain his own role in its recent history. He seemed to hold all the top jobs in the bergamot business. 'I am president of the new bergamot consortium that we set up three years ago,' he said, hurtling down the slip road and on to the motorway. 'And president of Unionberg, a bergamot producers' organization that I have set up.' Turning his left shoulder to the road and his face to me, like a man settling himself comfortably on a sofa, he went on to explain that the original bergamot consortium, established in 1934, had been stripped of most of its powers, but he was still its *commissario straodinario*, or special commissioner. These things are easier to say in the dark, but even this skeletal account of dismantling and rebuilding was ingrained with the sense of an industry in crisis.

To our right, the dark waters of the Strait of Messina lay as still as a small lake, the lights of Sicily a stone's throw away on its opposite shore. Pizzi looked across me towards the window in the passenger door. 'I travel all over the world,' he said, 'but I can find nowhere as beautiful as this.' 'This' is not Italy or Calabria, but the very small province of Reggio, where the history of bergamot began.

~

Pizzi was a generous host, and the restaurant he took me to in Reggio Calabria was a fine one. It was midweek, the place was almost empty and the signora front of house turned the full focus of her stern attention on us. She recited the menu like cannon fire, producing a litany of Calabrian fish dishes entirely outside my experience. Pizzi saw me hesitate and ordered for both of us. 'We don't like anchovies,' he said confidently.

He built the background to his story as the first *antipasti* were brought to the table. Many attempts have been made to grow bergamot in other places, but 90 per cent of the world's supply of essential oil of bergamot still comes from Calabria. The remaining 10 per cent is of an inferior quality because there is nowhere else in the world that can exactly reproduce the combination of rich soil and benign climate that the trees enjoy here. This inferior oil is extracted from fruit grown in Argentina, Brazil, Mali, Guinea, the Cameroons and the Ivory Coast.

There was a time, lasting a century or so, when Calabria's bergamot farmers were rich. In the 1960s a farmer with a hectare of bergamot groves in Calabria could expect to earn €50,000 (£42,500) a year in today's money. However, the next chapter in Pizzi's story traced the unfolding of a series of disasters. First came his chilling account of medical research in the 1970s that seemed to link bergamot oil to skin cancer, an idea desolate enough to dampen my appetite. It didn't matter that the results proved false; the damage to the reputation of bergamot oil had already been done, and the cosmetics and perfume industries had substituted genuine oil with a synthetic substance developed in the laboratories of northern Europe. As this substitute was very much cheaper than the real thing, many companies continued to use it after bergamot's innocence was proved.

More *antipasti* arrived, dish after dish as disaster followed disaster. Pure essential oil began to be adulterated by unscrupulous dealers, and when it was released on to the market it undermined bergamot's fragile reputation still further. And then there were the speculators whose games with the market made the price of oil fluctuate so wildly that farmers' incomes became almost entirely unpredictable. Some were willing to lurch on from year to year, but as tourism developed along the coast of southern Calabria, others preferred to grub out their trees and sell the land to developers willing to pay good prices and build hotels.

Then Pizzi's story jumped to the 1980s, and whenever I remember the first attempts made by the French to grow bergamot on the Ivory Coast, and the horror of Calabrian farmers when it looked as though the experiment would succeed, I remember them always in conjunction with an oval dish of stewed fish, rich with tomatoes and fiery with Calabria's famous chilli peppers. And still Pizzi talked, as a *primo* of pasta followed the *antipasti*, and the signora's disapproval of my unsatisfactory appetite grew ever darker and more profound. I redeemed myself momentarily by finding time to concentrate on the delicate and perfectly baked sea bream that was our *secondo*. With it Pizzi began on a tale of redemption in which he has played an ingenious role. Over the past three years he has spearheaded the establishment of the new Consorzio di Tutela del Bergamotto, a body with the power to control the quality of bergamot oil all over the world, and the legal right to test and withdraw inferior products on sale in Europe.

Since 1999, genuine Calabrian bergamot oil of a high quality has been distinguished by a DOP (*Denominazione di Origine Protetta*), a definition similar to but even more powerful than the DOC (*Denominazione di Origine Controllata*) given to wines from certain Italian regions. He has persuaded 75 per cent of bergamot growers in Calabria to join Unionberg, the private consortium set up to take over the old functions of the original *consorzio*. By joining forces, individual growers have more negotiating power, and dealers in bergamot oil find it more convenient to have a relationship with one organization instead of negotiating with several individual growers. The price of the oil is now fixed for an eight-year period with the dealers, and this has foiled the speculators' games and ensured growers a secure income.

Over coffee Pizzi explained that he was trying to improve conditions for growers still further by finding a bank willing to give a bridging loan to Unionberg each year. With this agreement in place, Unionberg could make a basic payment to them as soon as they delivered their fruit to the consortium. 'By that time they have already been waiting a year for some money,' he explained. The rest

of their profit would be paid to them later, after the essential oil had
been extracted and sold.

~

The following morning a brisk wind off the sea made palm trees in
the streets of Villa San Giovanni thrash about against the grey
clouds. I stood in the foyer of the hotel and gazed across the piazza
at the station. Every so often the top deck of a ferry appeared in
the sky above its roof, like an enormous and ungainly flying
machine. The morning loomed large and long, but then I remembered the *degusteria* offering bergamot tasting. I made my way out
into the blustery day and over the broken pavements. At first sight
the *degusteria* looked like little more than a lean-to, its doors and
windows festooned in padlocks. A security guard with a gun
prowled up and down in front of the bank next door. I took refuge
in a newsagent's on the other side of the road. 'Does the *degusteria*
ever open?' I asked. 'Always,' the newsagent said, and almost at once
a car drew up outside and the place sprang into life. Dodging the
juggernauts, I was beside the door as the padlocks came off.

'I'm forming a queue, Signora,' I said, as she fumbled through a
huge bunch of keys. By the time the door was open, I had introduced myself, and I already knew that she was called Sofie. I
explained my interest in bergamot and her face lit up. '*Allora, ci
scagliamo,*' she said. That was a word I didn't know, and when I asked
what it meant she paused for a moment and then replied, 'You know
how it feels when you lie down in the sun? Well, you might say, "*Che
sciallo!*"' She accompanied the word with an enormous and con-
tented sigh. I liked this promise of indulgence and relaxation, and so
I was glad to sit down at a small metal table inside the door. Beyond
its unimpressive exterior, the *degusteria* opened out into a pretty restaurant and bar. The shelves inside were lined with local products,
most of them preserved in bergamot syrup. There were hazelnuts
and almonds, pistachios and figs, oranges and even bergamot itself
preserved in its own syrup.

Bergamot is so acidic in its natural form that cooks ignored it for many centuries, and it was even categorized as a toxic substance. Pizzi fought for six years to have the toxic label removed. Eventually he managed to persuade the Department of Health to have the fruit and its peel analysed, and as a result the label has been suspended. I was glad to know that when Sofie returned to my table with fresh bread, cubes of local cheese and small dishes of bergamot marmalade; a jam made from Calabria's finest red onions, grown in Tropea on the Tyrrhenian coast; an angry jam made from local chillies and a glass of bergamot cordial to wash it all down. Then came figs and round chillies – like cherries with pyrotechnic intent – all of them suspended in an amber haze of bergamot syrup. I took things slowly, trying as I ate to put words around the taste of bergamot. Citrus was there, fresh, powerful and distinctive, but it was overlaid with bergamot's own, unmistakable and startlingly intense incense-like flavour.

~

Recently there has been something of a revolution in the use of bergamot, led in part by Professore Vittorio Caminiti. He is the owner of two *degusterie* and restaurants in Villa San Giovanni and another in Reggio Calabria. He founded the Accademia del Bergamotto, an association of growers and enthusiasts devoted to discovering new uses for bergamot and marketing bergamot products.

Caminiti's family founded one of the oldest hotels in the region, and he carried on the family tradition by becoming a hotelier himself. A few years ago he took a course in cooking and gastronomy, and since then has devoted himself to developing the bergamot-infused recipes that are served in his restaurants and *degusterie*. He has also written a surprising companion to Edward Lear's *Journals of a Landscape Painter in Southern Calabria and the Kingdom of Naples*. Caminiti is a great admirer of Lear, but in his opinion Lear showed a disappointing lack of interest in food.

Caminiti's companion guide is a booklet called *Viaggiando mangiando Mangiando viaggiando: Itinerario turistico-gastronomico dal 1847 a tavola con Edward Lear nella provincia di Reggio Calabria* ('Travelling Eating, Eating Travelling: A Gastronomic Itinerary for Tourists from 1847, Dining with Edward Lear in the Province of Reggio Calabria'). In it he retraces Lear's footsteps, amplifying his accounts of the meals he was offered in each place he stayed, giving detailed instructions on how they might have been prepared. 'Lear would undoubtedly have gone into ecstasy over bergamot sorbet,' Caminiti remarks, 'a velvet cream, cold, white, almost candied snow, leaving an aftertaste and perfume that are almost indescribable', although the recipe for this delicacy was actually invented in the twenty-first century by Fortunato Marino, an ice-cream maker in Reggio.

Professore Caminiti has developed several new recipes using bergamot. I saw a most impressive one when he invited me to join him at a wedding reception taking place at his hotel in Villa San Giovanni. I arrived as the celebrations were reaching a climax. Someone played a *zampogna*, the Calabrian bagpipe, and the older guests began to dance to its wheezing music. The cake was a huge and elaborate affair made from bergamot-infused sponge and cream. Caminiti called it a *Torta Nosside* and the recipe for it is in his book entitled *La Dolce Via*, published by the Provincia di Calabria.

TORTA ALLA BERGAMOTTO NOSSIDE

Caminiti assumes that you already have a three-layer sponge cake and bergamot-infused syrup, which is made by boiling the rind and juice of 1 bergamot in 250ml water before adding 100g caster sugar and simmering gently for 15 minutes. The recipe below is for a bergamot pastry cream to sandwich and ice the cake, plus caramelized bergamot slices to decorate it.

4 whole eggs
15g softened butter
375g caster sugar
60g plain flour

> *2 bergamots*
> *500ml milk*
> *250ml very lightly whipped double cream*

- Beat the egg yolks, softened butter and 175g sugar together in a bowl with a wooden spoon, then stir in the flour.
- Place the rind of 1 bergamot in a saucepan with 500ml milk. Bring to simmering point, then remove from the heat and leave in a very low oven for an hour to infuse.
- Strain the infused milk into the bowl with the egg mixture, then tip everything into a heavy saucepan and cook over a low heat, stirring all the time, until the mixture thickens. Keep cooking for about 15 minutes, or until it reaches the consistency of a thick white sauce. Add more milk if it gets too thick. Allow to cool.
- Divide the pastry cream into three portions. Take a layer of sponge, sprinkle it with bergamot syrup and cover with a third of the pastry cream. Add the second layer of sponge, sprinkle it with syrup and cover with another portion of pastry cream. Top with the third layer of sponge, sprinkled with more syrup.
- Refrigerate for at least 30 minutes before mixing the remaining pastry cream with the very lightly whipped double cream and using it to cover the chilled cake.
- Cut the remaining bergamot into slices, then place it in a pan and cover with 200g caster sugar. Cook until the foaming sugar has turned light brown and just started to smoke. Tip the bergamot slices and caramelized sugar out at once on to baking parchment. Use the caramelized fruit and shards of caramelized sugar to decorate the cake.

Caminiti's other great contribution to the history and culture of bergamot is a museum tucked away in a narrow street in the centre of Reggio Calabria. The entrance takes you straight into a bar, then a restaurant. When Ezio Pizzi took me there, I soon found myself sliding across the huge, shiny dance floor of a disco. The museum was on the far side of the disco, and inside we found an entire wall lined with examples that plot the evolution of the *macchina calabrese*, the machine invented by Nicola Barillà in 1844 to revolutionize the laborious extraction of bergamot oil. When Pizzi began enthusiastically turning the metal handle of the *macchina*, the museum guide stepped abruptly aside to avoid his flailing arm. His efforts caused a large wooden wheel on top of the machine to revolve and this turned a spindle with a pair of metal cups at its base. If only we'd had some fruit, I'm sure Pizzi would have seen the extraction process through from beginning to end.

When we had finished in the museum, we were invited to sample a bergamot-infused cocktail at the disco bar. We took our places on two bar stools that were covered in slippery white leather. 'This is a cocktail that I have invented for strong men,' the barman said, placing a large vermouth glass filled with ice on the bar in front of us. 'I have collected some stones from the riverbed and marinated them in a mixture of vermouth and bergamot oil.' And indeed, there they were, rolling about like specimens in a jar. He tipped the ice from the glass and replaced it with three stones that he removed from the jar with a pair of oversized tweezers. 'Almost there,' he said, and filled the glass to the brim with gin. I waited expectantly, but he picked it up and handed it to Pizzi. 'My cocktail for a strong man,' he said. Pizzi took a sip, and then another. 'Now I will prepare my ladies' cocktail,' the barman said, reaching for a tall glass. He began by filling it with ice. 'Next I add fig molasses and brown sugar.' The sugar was followed by a teaspoon or two of vermouth and topped up with something bright and fizzy. As a final insult, he thrust two colourful straws deep into the ice and slid it across the bar in my direction. While I took a first sip, Pizzi began a tentative experiment with the swivelling capacity of his bar stool. It spun out of control and our knees collided. 'Does he have a cocktail for the mature woman?' I whispered.

We had a delicious dinner made up of many tiny bowls of local specialities in the restaurant, and afterwards Pizzi paused for an espresso at the bar. I didn't want one and the barman was immediately racked with concern for my digestion. 'Don't worry,' Pizzi said, when I turned down a *digestivo* as well, 'she had a bergamot sorbet.' That placated him and we were free to leave.

~

In Bova, a small town close to the tip of Italy, a family of the same name has been extracting bergamot oil for four generations. Annunziato Bova had already stopped work for the day when Pizzi took me to meet him, but the scent of bergamot still hung in the air. The shining metal machinery was as clean as a Swiss laboratory, and Bova's tiny son was engaging with the family business by using the polished fruit hopper as a slide. Bova led me through the extraction process stage by stage. When the fruit arrives, he explained, it's washed in the hopper and then channelled into a rotating chamber lined with graters. This is a highly skilled business and Bova must check the graters every few minutes, sometimes adjusting them by less than a millimetre to ensure that they remove only enough peel to release the oil. During this process the fruit is rinsed continuously with powerful jets of water. As the water, rind and oil run out of the chamber in a constant stream they are carefully sieved. The oil and water emulsion is then put into a centrifuge for separation.

When I visited it was early November and the oil that dripped from a narrow pipe after that day's extraction was shiny and dark. This was green gold, the most valuable oil of all, and it would be used only in the highly lucrative context of the perfume industry. By the time the harvest ended in December, the fruit would have turned yellow and the golden oil extracted from it would be good only for flavouring cakes and confectionery. But whether a harvest is at its beginning or its end, extraction is always a painfully slow process, and 100 kilos of fresh fruit make only 500 grams of essential oil.

Bergamot's prime cut is its oil, but after that has been extracted the fruit is pressed for its juice. This is rich in antioxidants and recent research has shown that it can be added to other fruit juices to enhance their antioxidant activity. It can also provide the ascorbic acid that acts as a preservative and is generally added to juices in synthetic form during industrial processing. After the juice is exracted, the crushed fruit is left to dry before it's fed to cattle. Francesco Calabrò, the doctor who first noticed bergamot's healing powers, remarked that the dried rinds could also be used as fuel during the winter, or hollowed out and transformed into the strange, elaborately carved tobacco pouches that can still be seen in the bergamot museum. They are said to have kept tobacco moist and imbued it with the fresh flavour of the fruit.

Immediately after extraction the oil is stored in barrels in the small room on the other side of the Bova's narrow yard. Bova dipped an enormous pipette into a barrel and drew out just enough dark green oil to fill a small bottle. Like a sommelier with a glassful of something especially fine, he lifted it to his nose and inhaled deeply. 'This is a good year,' he said, 'a very good year.' He handed the open bottle over and now Pizzi inhaled. 'It is good,' he agreed, 'exceptional.' And then it was time for my first experience of bergamot oil. I couldn't make comparisons, but what I smelt was wonderful. It had the brightness and clarity of sunshine, overlaid with a deep, incense-like quality. It was the essence of citrus, and yet it didn't resemble the scent of the lemons or sour oranges that are its ancestors.

Giorgio Gallesio, the early-nineteenth-century citrus expert whose villa I had seen on the hill above the Parodi smallholding in Liguria, had his own explanation for the individuality of bergamot's magnificent scent. He said, 'We are now convinced that, with the same principles differently combined, Nature diversifies greatly her products, and consequently it is very probable that the combination of the odorous principles of the lemon with those of the orange may give a result still more exquisite than either alone.'[10] We stood for a moment in silence before Bova turned to me and said, 'I'm really, really proud of what I do.' And then, gloriously, unexpectedly, he pressed that precious bottle of green gold into my hand.

Unique Harvest

On the Riviera dei Cedri

꩜

The unique qualities of Calabria's bergamot are matched only by that of a citron found in the north of the region. Although the citron has more varieties than almost any other kind of citrus, Calabria is the only place in the world that produces the *Cedro liscio di Diamante*, the smooth-skinned Diamante citron that takes its name from a small town in the north of Calabria. For Lubavitcher Jews, a Hasidic movement of Orthodox Judaism, the Diamante citron is one of life's essentials. They believe that Moses sent a messenger to Calabria to collect an example of the local variety of citron. This enabled him to show the Jews the fruit they should use during the celebration of Sukkoth (the Festival of Tabernacles), an autumn festival of thanksgiving for the harvest and the survival of their ancestors as they wandered through the desert for forty years. There are at least twelve distinct varieties of citron, or *esrog* as it is in Hebrew, acceptable for use during Sukkoth. *Esrogim* are also cultivated in Israel, Morocco, Mexico and California, but Menachem Mendel Schneerson (1902–94), leader of the Lubavitcher movement, taught his followers to favour the *Cedro liscio di Diamante*, and so for Lubavitchers only citrons grown on a forty-kilometre stretch of the northern Calabrian coast between Tortora and Belvedere Marittima will do.

Citrons were the first species of citrus tree to reach Europe and they have been cultivated in Calabria ever since the Jews arrived there in the second century AD. Today, citrus is an enormous genus with an incalculable and ever-expanding membership, and yet originally the citron, the pomelo and the mandarin were the only species of citrus in the world. When you mention citrons, people

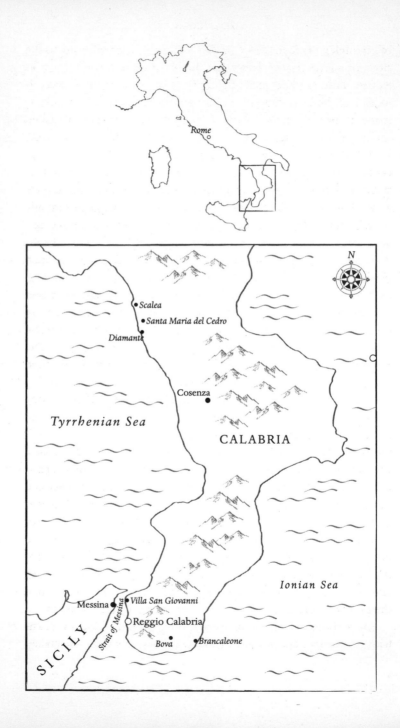

often think you mean *Citrus*, or they assume you are referring to the *citron* of a French *citron pressé*. But this is not *Citrus limon*, it is *Citrus medica*, a much more ancient and primitive object than a lemon. It looks like the beginning of an idea about fruit, a rough prototype made at an early stage of the design process, a crude unfinished thing, a dinosaur that evaded extinction, a Neanderthal on a tree. And if you did try to *presser* a citron, the results would be disappointing because its skin and bright-white pith are so thick that there is space inside for only a spoonful or two of flesh, scant juice and a few seeds. This extraordinary carapace changes gradually from luminous green to a deep golden yellow as it ripens. It is saturated in essential oil and its surface is sculpted into ridges like a downland landscape, or raised in terrible carbuncles. It has the open pores of an alcoholic's nose and it exudes a perfume so powerful that it can engulf the ground floor of a house, moving from room to room in penetrating swathes. It smells stronger, wilder and more exotic than a lemon, like the lemon's big brother, like Mediterranean heat cut through with sweet violets and something spicy. The essential oil stored in the citron's skin contains a compound called beta-ionone, and it's this that contributes violets to the aroma.

The citron's history began in remote and sheltered valleys among the foothills of the Himalayas in Assam. Those wild citrons were straggling and bush-like, with long thorns and fruit so heavy it pulled their feeble branches to the ground. Assam offered the ideal climate for such tender, thirsty trees. It has one of the highest rainfalls in the world, most of it compressed into four months between June and September, and even in the cold season between October and February the temperature doesn't drop below eight degrees Celsius. So the citron is spoilt by nature, an aristocratic infant raised in a perfectly appointed nursery, fragile, fussy and ill equipped for the harsher reality of a wider world. I wouldn't be surprised if an atavistic longing for this ancient Eden had coursed like sap through the branches and trunks of citrons ever since they arrived in Italy.

For a thousand years the citron was the only kind of citrus fruit in Europe, and it did not lose its monopoly over the Italian peninsula until Arabic invaders brought lemons and sour oranges to Sicily. Left to its own devices, the Diamante citron would grow like a huge bush, its branches dragged down by the weight of fruit that would lie rotting on the ground. However, a citron grove in Calabria is a highly disciplined place. Wires run along the lines of trees at about chest height and the citrons' trailing branches are trained along them. If the wires were higher it would be easier for the pickers, but raising the tree's limbs too high results in a poor crop. Consequently, anyone working with citrons must treat them like kings or deities and approach them only on their knees. From a distance, the field looks more like a vineyard than a citrus grove. Get a bit closer and it's clear that the wires support the juggernaut of the citrus world. Its long branches are freighted with large oval leaves, lethally sharp thorns, richly scented flowers and an extraordinary load of diamond-shaped fruit with undulating, iridescent green skin. Some citron varieties are sweet, but the fruit of the *liscio di Diamante* is bitter and when it's mature it can be up to thirty centimetres in length and a kilo in weight, with skin so thick it makes up 70 per cent of the fruit's volume.

The citron began its journey towards Calabria by migrating slowly into China and across India. The climate it encountered outside Assam was so much hotter, drier and more challenging that it couldn't survive without human help, but what possible appeal could it make to a farmer? Would you choose to eat its fruit? Not really. Was its wood good for burning? Not very. Was it useful for building? Not at all. Could anyone find shade beneath its branches? Certainly not. Did it at least live for a long time? No. So it was a practical failure, and yet there was something miraculous about it that could not be ignored. It had an almost supernatural ability to bear a full cargo of beautiful flowers and enormous golden fruit simultaneously throughout the year. Everything about it was scented – its pale waxy flowers, its dark green leaves, its fruit and even the wood itself – and like a glamorous woman, it was constantly surrounded by a miasma of perfumed air. Finally, the fruit seemed eternal, neither rotting nor falling from the tree. Although it had no obvious practical use, the tree's mysterious habits gave it a powerful and peculiar appeal, so that people seem always to have felt compelled to cultivate it, imbue it with symbolic significance, paint its portrait and include it in ancient stories.

In India the citron was referred to as *jambila* in *Vajasaneyi Samhita*, a collection of sacred texts written in Sanskrit before 800 BC and associated with Kuvera, the golden-skinned, pot-bellied god of wealth, who looks rather citron-like himself.[1] He holds the fruit like a glowing orb in his 'boon-giving' right hand, while his left hand grasps the collar of a mongoose that spits jewels into his lap. As a result of this association, and perhaps because of its gigantic size and golden colour, the citron became a symbol for prosperity. It appears again in a faded fragment of a painting on the ceiling of an ancient cave near Ajanta in central India. The cave is part of a complex of rock temples and hermitages, most of them made in the fifth century AD. Its roof is decorated with a *trompe l'oeil* painting of a coffered wooden ceiling. Panels illustrating the life of Buddha alternate with fruit and flowers, and here the fruit's bulky, square-shouldered profile is unmistakable.[2]

The citron spread gradually from India into Persia, its fruit stowed deep inside the saddlebags of merchants moving along the caravan routes that ran from the Punjab in upper India through Afghanistan to Persia and Mesopotamia (between the Tigris and Euphrates rivers in Iraq), or flashing gold among the cargoes of boats carried on monsoon winds from the west coast of India to Oman, before being taken overland to Iraq and then Iran. Citrons travelled well: they were slow to decay and their seeds were protected by the fruit's enormous carapace of pith and peel. The trees were fully acclimatized in Persia and Media (north-west Iran) by the fourth century BC, when Alexander the Great came storming through with his armies and a vast retinue of scientific experts. The scientists were commissioned by Alexander to record every aspect of the flora and fauna, geography, people, mineral deposits and infrastructure of the regions they passed through in the wake of his armies.[3] They were on the lookout for useful trees or crops that might be acclimatized in Macedonia and Greece, and in the course of their travels they amassed a wealth of detailed practical information. And of course they noticed the citron, a tiny tree with fruit like a giant's tears, and they called it the apple of Media. They cross-questioned the farmers who grew it, finding out all they could about its uses and cultivation.

After Alexander the Great's death in 323 BC all the material gathered by his retinue of scientists was placed in the state archive in Babylon. The information on plants and trees was eventually given to Theophrastus, the Greek philosopher, author and natural scientist. He used it as a source of material for his *Enquiry into Plants* of 310 BC, the first text ever written to present botany as a science in its own right.[4] His main interest was in the classification of plants and trees, but he also wrote about their cultivation and use. In Book IV, Chapter IV, 'The Trees and Plants Special to Asia', he wrote, 'And in general the lands of the East and South have peculiar plants, just as they have peculiar animals.' Cue the citron, which he calls a *melon Medicon* ('Median fruit') or *melon Persicon* ('Persian fruit'). Persian gardeners had already observed that citron flowers without a pistil were sterile. They reported their observations to Alexander's scientists, who appreciated the scientific importance of this discovery, and the role of the pistil in the process of fertilization and fruit production was acknowledged for the first time.

～

Theophrastus never pretended to have seen a citron himself. In fact he made it quite clear that his account was third- or at best second-hand by using the stock phrase 'so it is said' at frequent intervals. From Theophrastus we can deduce that the inhabitants of Persia and Media thought the citron inedible but found its juice a wonderful antidote to poisoning, because when mixed with wine it could be relied on to induce vomiting. They also used it as a giant mothball that could be placed among clothes to repel insects – a good idea, as we now know that citron peel contains limonoids, which are natural insecticides – and to perfume them. This is still an excellent use for a citron, as I inadvertently discovered on a return trip from Calabria. I'd taken an old suitcase with me, the kind that is always on its final trip and yet manages to be beside you, like a disreputable friend, in the lobby of the next hotel. It didn't have much in it when I arrived, but by the time I left it was stuffed full of citrons. Their weight was extraordinary, the case's fragile wheels buckled beneath it and when I left the hotel I had no choice but to accept the assistance of a disapproving porter. Together we hauled the case across the car park and wrestled it, like a drunk or a man resisting arrest, into my car. When I got home and opened it, the room filled with the smell of the fruit, a combination of spice and sweet violets that had entered every fibre of my muddy clothes. I'd packed some books as well, and to this day their pages are impregnated, so I only have to open one of them to catch a ghostly scent of citrons.

The Romans also used citrons to scent the air of a room and they are still used for this purpose in China. I once lined some citrons up on the dresser in our kitchen, and they sat there for weeks, looking like beautiful ornaments and gently filling the house with their perfume. But the kitchen's warmth eventually took its toll, drying and diminishing them. Suddenly their skins stopped fitting, as if they were coats borrowed from someone bigger, and their folded contours began to deepen. In the end they looked like a row of wrinkled sages, but a rich, dark scent still surrounded them.

The idea of the citron reached Europe in Theophrastus's treatise in 310 BC, but it wasn't until the fourth century AD that the Roman author Palladius dedicated a chapter, *De citreo*, to citrons in his *Opus agriculturae* ('On Agriculture, in Fourteen Books'), giving us the very first account of citrons growing on Italian soil. Until then Roman authors, such as Virgil in 30 BC or Pliny in AD 77, were content to reproduce the small, undigested and by now over-familiar bolus of information about citrons passed on by Theophrastus almost 400 years earlier. And what else could they do if they had never seen a citron for themselves, let alone touched or tasted one? Palladius could finally write in refreshing detail about the trees growing in his own citron groves on the island of Sardinia and near Naples, and he could also offer lots of first-hand expert advice. He stressed that citron trees 'love a light earth, a warm climate and continual humidity'. In order to plant citron seeds he tells us to dig the soil to a depth of two feet and incorporate ashes into it. We should divide the seedbed with small earthen berms that will channel and retain water, sow the seeds in threes and water them daily with tepid water. When the seeds grow into trees, we're to plant gourds under them because citrons benefit greatly from the ash produced by burning gourd vines at the end of summer. Their fruit must be gathered on a moonless night, covered in leafy branches and stored 'in a secret place'. Some people keep each fruit in 'a separate vessel', others store it in the shade and cover it in gypsum, but 'most people keep them in sawdust, or in cut straw or chaff'.

Classical writers used a variety of different names to refer to the citron. The Greeks began by adopting the names used by Theophrastus, and when Virgil wrote about it he called it 'the Median apple tree'. In the middle of the first century AD the Romans began calling citrons *citreum* and *cedrium*, names also used for the cedar tree. This confusion may have been due to the fact that both citrons and cedars had powerfully aromatic wood. When Linnaeus devised a modern system of nomenclature for plants and animals in the eighteenth century, he made *Citrus* the name for the entire genus. His use of the word *medica* makes the citron sound more like a medicine than a fruit, but it derives from Media, the kingdom of the Medes, where its presence was first recorded.

∼

Citrons were thought inedible by the ancient Greeks, but by the first century AD Plutarch was including them among 'many substances which in the past people would neither taste nor eat, that are considered today as very agreeable'.[5] The citron found its way into *De re coquinaria*, a collection of recipes published under the name of Apicius. No cook himself, Apicius was famous among his contemporaries only for his wealth and his exceptional greed. The recipes were actually those of slave cooks.[6] Among them is a power-ful sauce made by chopping up citron peel with mint and fennel and mixing it with broth. I once tried this recipe myself, using a beauti-ful citron with undulating golden skin that I had brought home with me from a garden in Tuscany. I was unwilling to sacrifice the whole fruit to this dubious cause and so I just cut off its pointed tip. As I picked up the nugget of dazzling-white pith covered in gold skin, the air was suddenly drenched with perfume. By cutting the rind I had released essential oils from the pores just beneath its sur-face, and the smell of violets that comes from a citron's outer skin had been replaced by a scent reminiscent of crushed geranium leaves in the sun mixed, as ever, with warm but indefinable spices. This wilder, more concentrated and exotic perfume is the essence of citron, its truest and deepest scent.

I diced peel and pith together with fresh mint and fennel. There was stock in the freezer, but it smelt rank in the face of the citron's almost antiseptically fresh perfume and so I made up some bouillon from a jar. 'And now what?' my companion asked. 'Are you going to cook it?' Panic set in. There had been something altogether too cas-ual, too spontaneous about the whole event, and now I felt like a surgeon afflicted by amnesia halfway through an operation. I couldn't remember anything in Apicius about heating the ingredi-ents, and yet why should the simple addition of broth turn that chewy pile of herbs and raw citron peel into a sauce? I dashed back to my desk and sifted through the papers there. I found nothing to suggest cooking the sauce, but we did it anyway, hoping for some kind of melding and softening of the ingredients, as I didn't want the citron to have been sacrificed in vain. There was no melding or softening. It was omelette for supper, or rather a *frittata* full of potatoes, peppers and cheese. We put pools of the sauce on the edges of our plates and made timid forays into them with forkfuls of *frittata*. The flavour was astoundingly powerful, almost as if we were attempting to eat incense. It made a sad sort of supper, a bemused wake for that beautiful citrus monster, now irretrievably defaced.

Citrons are consumed in the imaginary dinner described by Athanaeus in *Deipnosophistae*, or 'Scholars at the Dinner Table', but not for pleasure. Athanaeus, a Greek who moved to Rome, wrote *Deipnosophistae* in the third century AD. In it he gathered all his heroes, alive and dead, around an imaginary table for a meal that seemed to last for several days. When they spoke of the citron, their principal interest was in its use as a defence against – rather than an antidote to – poison. The focus of this discussion was a story set in the imperial province of Egypt, where a couple of convicts were being led through the streets to certain death in the snake pit. A shopkeeper's wife standing in the street took pity on them and, taking a bite from the citron she happened to be holding, she offered the mouthful to one of the men. He ate it, the sad procession continued on its way and before long he was tossed into a pit seething with poisonous snakes. The other prisoners were killed by the snakes' venom but the man who had eaten citron was unharmed and the next day the governor of the jail decided to conduct an experiment. He ordered one prisoner to eat a citron and gave another nothing at all. Afterwards they were sent to the snake pit, where once again the man with a bellyful of citron survived unharmed and the other died instantly.

No wonder the scholars at the dinner table were so keen to have a recipe for citron that 'when eaten before any food, dry or liquid, is an antidote to every poisonous ingredient'. The method for its preparation was very simple. You boiled a whole citron, seeds and all, in Attic honey until it dissolved. 'And anyone who takes two or three fingers of it in the morning will not be harmed by poison in any way.' The imaginary scholars were so impressed by this that they gobbled up the imaginary citrons on their imagined table 'as though they had not touched any food or drink before'.

Citrons also had more specific medicinal applications in ancient Rome. Galen, another Greek, moved to Rome during the second century and practised as a doctor in the gladiator school before becoming private physician to Emperor Marcus Aurelius. He recommended chewing citrons as an antidote to stomach cramps and prescribed them for heart problems 'on account of their innate heat'. And at about the same time Scribonius Largus, doctor to the court of Emperor Claudius, suggested cooking a citron in vinegar, mashing it together with alum and myrrh and applying it to the 'ruddy swelling' caused by gout. This approach was very conservative in comparison to his other cure for gout. Perhaps it was only when the citron poultice didn't work that Scribonius Largus prescribed treatment by torpedo ray fish, a very early form of electro-analgesia, with the fish providing the electric current. The patient was made to stand in shallow water on the edge of the beach, a torpedo ray firmly trapped beneath his foot. The ray, or perhaps a succession of rays, gave his foot repeated electric shocks, eventually numbing it so completely that he could no longer feel the pain.

~

In ancient Greece and Rome the citron served as a symbol, just as it had in India. This time it was associated with Herakles, or Hercules as he was to the Romans, and the eleventh of his twelve labours, when he stole the golden apples from the Gardens of the Hesperides. 'Apple' was a generic word for fruit in ancient Greece and Rome, and the golden apples were originally depicted as quinces, as they were in the frieze in the Temple of Zeus, built in Olympia in 450 BC.[7] There's certainly something mysterious about a quince, with its dusky, moleskin pelt, but in the first century AD Juba II, King of Mauretania, referred to citrons rather than quinces as 'the apples of Hesperia'.[8] In the second century AD, when the labours of Hercules were portrayed on a series of bronze drachma coins struck during the reign of Emperor Antoninus Pius, Hercules is shown picking citron-shaped fruit from the tree.[9]

Citrons, or sometimes oranges, were still present in images of the Gardens of the Hesperides in the sixteenth century, when the myth became a rich source of iconography for the Italian Renaissance garden. Over the years I have come to recognize Hercules as you might a colleague working in the same organization as you, but in a very different role. I'm usually moving though the garden with groups of visitors, while he's generally intent on mashing a hydra to pulp with his club, wrestling a lion or straining every muscle to haul Antaeus off the ground. We've not had much in common in the past, but once I began to learn about citrons, everything was different. Our shared interest transformed him from a mythical figure into someone almost real, a huge, shock-headed giant of a man with impossibly broad shoulders and gleaming skin. He's ingenious, cunning, fearless, and has no qualms about violence. The theft of the golden apples depended on every one of these qualities. Without ingenuity he'd never have found the Gardens of the Hesperides, because nobody seemed able or willing to tell him where they were. Thanks to skill and a violent streak, he was able to shoot an arrow over the garden wall and kill Ladon, the many-headed dragon guarding the tree. And without cunning, he couldn't have persuaded Atlas to steal the apples for him, or duped him into taking the weight of the celestial globe back on to his own shoulders, just when he thought he'd escaped that back-breaking burden for ever.

By the sixteenth century several new uses had been found for citrons in Italy. In Cristoforo Messisbugo's *Libro novo nel qual s'insegna a far d'ogni sorte di vivanda* ('New Book in Which the Preparation of Every Kind of Food is Taught') there is a detailed account of a banquet given in Ferrara on 20 May 1529 by nineteen-year-old Ippolito d'Este. The table was set out beneath trees in a twilit garden. Three white cloths covered it, one on top of the other, and both the table and the branches above it were decorated with swags of flowers. Each place was flanked by a folded napkin, a knife, a bread roll and a citron, its skin intricately engraved with the initials and coat of arms of the guest.

Pietro Andrea Mattioli's *Commentaries* (1544) on the *materia medica* of Dioscorides recommended using the essential oil of citrons to preserve dead bodies and suggested rubbing the genitals with it before making love to prohibit conception. Vincenzo Corrado, whose recipe book *Il Cuoco galante* was published in 1778, found several culinary uses for citrons. He gives recipes for a variety of citron-flavoured drinks, including a liqueur that he names *Labro rubino* or Ruby Lips, which gets its lip-staining qualities from the addition of sandalwood. There are also recipes for marmalade, candied citron peel and a citron-scented vinegar.

VINCENZO CORRADO'S RUBY LIPS

Take two pints of spirits distilled from wine and add to it half an ounce of ground cinnamon, half an ounce of ground sandalwood, four ground cloves and the peel of a citron. After four days, sieve and add two pounds of sugar dissolved in a pint of hot water. Allow to settle before filtering.

The citron could fall in and out of fashion in the kitchen, but for observant Jews *esrogim* occupied a position from which they could never be ousted. Jewish exiles first saw citrons during their captivity in Babylon after the fall of the Temple in 586 BC, and when the Persians conquered Babylon in 539 BC and issued a decree allowing them to return to their homelands, the citron tree made the journey with them from Babylon to Palestine.[10]

In Leviticus, Chapter 23, Verse 40, the Jews are given instructions for the preparation of Sukkoth: 'On the first day you shall take the fruit of the goodly tree, branches of palm trees, foliage of leafy trees, and willows of the brook and you shall rejoice before the Lord your God for seven days.' They interpret 'foliage of leafy trees' as the myrtle and 'the fruit of the goodly tree' as *esrogim*. On each day of the festival except *Shabbat* – Saturday, or the Sabbath – the *esrog* and the branches (or *lulav*) are blessed and then shaken together in all six directions, left and right, up and down, forwards and backwards. This part of the ceremony is called 'the taking of the Four Kinds'. The Four Kinds represent the different types and personalities of Jews that make up the community of Israel. The first are Jews who study the Bible and follow its commandments. They are represented by the citron, because it is both perfumed and edible. Next come those who follow the law but neglect Bible study. The palm, which bears edible fruit but has no perfume, is their representative. Then there are those who study but don't practise, and like myrtle they are perfumed but bear no fruit. The liberal Jews who neither believe in God nor follow his commandments are like the willow, which is pretty but can offer neither perfume nor fruit. In the eyes of God all are equal, and the shaking of the citron together with the three kinds of branches is an acknowledgement of the unity of Jews all over the world.

The *esrog* became a symbol of religious and national identity for the Jewish people, and from the first century BC the fruit's unmistakable profile, part fir cone, part pineapple, began to appear on frescoes, mosaics, tombs, inscriptions and ritual objects in Palestine. At times, citrons played a part in acts of rebellion, such as the historic event reported by Flavius Josephus in *The Antiquities of the Jews*. Alexander Jannaeus, Sadducee priest and King of Judea from 103 to 76 BC, was standing at the altar in the Temple celebrating Sukkoth when his own people rebelled against him by pelting him with the citrons held so conveniently in their left hands. And during the revolt against the Roman Empire between AD 66 and 70 they minted coins with an *esrog* in place of Nero's head.

Defeat was swiftly followed by the destruction of the Second Temple in Jerusalem. This combination of events triggered a mass exodus, so that new Jewish communities soon sprang up all over the Mediterranean. Citrons were a prerequisite for the rituals surrounding Sukkoth, and Jewish farmers and gardeners in Egypt, Asia Minor, Syria, the Aegean islands, Greece, North Africa and Calabria honed the skills required for propagating citrons from seed, nurturing the young trees, pruning them, harvesting and storing their fruit. They were soon renowned for their horticultural expertise, and over the centuries they extended their skills to the cultivation of other kinds of citrus, such as the lemons and oranges that found their way to the Mediterranean with the Arabs in the ninth century. Consequently Calabria and many other places settled by the Jews also became centres for commercial citrus cultivation.[11]

Until the sixteenth century, Hebrew would have been a familiar language in Calabria. Jews settled there first in about AD 70, and in the thirteenth century their presence was actively encouraged by Frederick II, the most powerful Holy Roman Emperor of the Middle Ages. Under Frederick's rule, Sicily and southern Italy became one of the most civilized kingdoms of the age. This sophistication was due in part to Frederick's religious tolerance. Under his rule the town of Reggio Calabria was an important centre for Jewish life, and Calabria benefited from the new ideas, inventions and goods brought there from all over the world by Jewish merchants and travellers.

Some people in Calabria still look back on Frederick's reign as a golden age and attribute the prosperity of his kingdom in part to the creativity and business acumen of the Jewish population. But from 1495 the region was ruled by Spain, as a part of the Kingdom of the Two Sicilies, comprising Naples and everything to the south of it. Under the Spanish, Calabria experienced nothing but poverty, famine, disease, high taxes and general misrule, and during the Spanish Inquisition of the mid-sixteenth century the Jews were expelled. Calabria was considered a reliable source of *esrogim*, and Jewish customers were content to buy the fruit unseen, have it sent north from Calabria and shipped out through Genoa and Sanremo. However, when the Jews of southern Italy were expelled or forced to convert to Christianity, Jewish communities in northern Europe began to doubt the purity of the *esrogim* grown for them by Gentile farmers.[12] Leviticus, Chapter 19, Verse 19, forbids cross-breeding in domestic animals, planting different kinds of seed in the same field, or wearing clothes made from two kinds of material. In rabbinic law this translates into a prohibition against the grafting of one species of tree on to another, a practice particularly common in the cultivation of citrus. This cast doubt on all citrons produced in traditional areas of cultivation in the Mediterranean.

Although it is easy to tell if a tree has been grafted, because the junction always remains visible on the trunk, this did not solve the danger of fruit being picked from an ungrafted tree grown from the cutting taken from a grafted tree. *Esrogim* picked from a grafted citron tree were considered particularly abhorrent because they presented the danger of a forbidden object being used for a sacred purpose.[13]

Anxiety created an opening for Jewish citron merchants who could reassure their customers by supervising the cultivation and harvesting of their goods. These days rabbis and merchants fly into Lamezia, the small airport serving northern Calabria, in time for the *esrog* harvest every year. Many have long side-locks and beards, and some of them defy the heat by wearing fur hats and long black overcoats. They make an eye-catching sight among holiday crowds at the airport.

~

To reach Calabria's citron groves you drive north from Lamezia, on empty roads, with the sun in your eyes and the Tyrrhenian Sea never far from your left shoulder. Santa Maria del Cedro, Diamante and Scalea are the principal towns in the area that the tourist board has rechristened Riviera dei Cedri. Santa Maria was originally called Cipollina, an unfortunate name because it is the same as the word for a small onion. In 1956 the local priest, tired of hearing his parishioners teased, had the name of the town changed to Santa Maria. He meant well, but unfortunately too many towns in Italy go by that name and its inhabitants soon had a new problem: their post constantly went astray. Ever inventive, the priest changed the name once again to Santa Maria del Cedro, St Mary of the Citron, an appropriate title for a town set in the landscape that has produced citrons for more than 2,000 years.

Over the centuries an extraordinary number of different rulers and invaders have settled in this area of Calabria. Once part of the Roman Empire, it has also been governed by Normans, invaded by Arabs and ruled by the Spanish. Each of these cultures has left its imprint and on the Riviera dei Cedri they speak *Cosentino*, a dialect bearing some resemblance to Neapolitan and very little to any form of Italian I can understand. Further south it's *Calabro*, which has closer links to Sicilian, and at the end of the twentieth century the inhabitants of many of the villages around Cosenza still spoke *Arbëresh*, a language derived from the *Tosk* dialect in southern Albania, which they considered the purest form of Albanian.

Citrons are tender, difficult trees, intolerant of wind, cold or drought. Fortunately there are mountains behind the citron groves on the Riviera dei Cedri and they contain and concentrate both the warmth emanating from the sea in winter and the cooler, maritime microclimate of summer, so that temperatures rarely fall below five degrees Celsius or rise above thirty. The citron's wood is unusually brittle and this makes it especially vulnerable to wind. Luckily the mountains also shield the trees from gales blowing off the Ionian Sea in winter. And warmer wet winds that blow off the Tyrrhenian Sea in winter collide with the mountains and drop an abundance of rain on the coast. In summer much of Calabria suffers from drought, but even in August the River Lao continues to hurtle down the mountains to the Riviera dei Cedri, before rushing on towards the sea. The last curve in its course sets it parallel to the coast. Cast yourself into it, as I did when I was last there, and it carries you out to sea, propelling you rapidly through shallow water over bright gravel, spinning you among twigs and leaves, pushing you like any other piece of flotsam into the tranquil, turquoise Tyrrhenian. But for all its eagerness to arrive, the river is slow to throw in its lot with the sea, and you swim through layers of cold, fresh water that hang suspended over the Tyrrhenian's warm depths. And if you've ever asked yourself, 'Am I too old to wear a bikini?' you should hurry to that grey shingle beach, where everyone is brown and looking beautiful, whatever their age or shape.

～

There are two citron harvests in Calabria. The first is in August, when most of the fruit is still small enough to grasp in the palm of

your hand, just as the Jews do when they carry it into the synagogue. All of the fruit from this summer harvest is used during
Sukkoth. Only a small proportion of the crop is of a high enough
quality to pick for this purpose, and the perfect fruit of this limited
harvest fetches the highest price. The second harvest is in November, when all the remaining citrons are as heavy as treasure, and as
big and sleek as overfed puppies.

The rabbis and citron merchants remain in Calabria for a month
to supervise the *esrog* harvest and buy the *esrogim* used during Sukkoth in Lubavitcher communities all over the world. The exact date
of their arrival varies each year according to the weather and the
maturity of the fruit. My earliest attempt to witness the *esrog* harvest failed because I relied on a local tourist office to tell me when
the rabbis and merchants were due to arrive. 'Nothing could be simpler,' said a hoarse voice on the other end of the phone. 'We'll
just call and tell you when the Jews are coming.' July faded into
August and no call came. I rang back a few times but there was no
sign, it seemed, of 'the Jews'. The two of us were like twitchers, I
thought, desperately scanning the horizon for rare migrants.

As the end of August approached I lost patience and began to
ring other contacts. 'Would you like the number of one of the directors of the Consorzio del Cedro di Calabria?' someone asked. The
citron consortium controls every aspect of cultivation in Calabria. It
was established in 2000 to market citrons and citron products and
promote their cultivation and the culture associated with them and
the area in which they are grown. I was given a series of phone numbers for Antonio Miceli, one of the consortium's directors. Miceli
never seems to stay in one place for long. He's a man in a hurry, with
multiple businesses and not enough time to run them. When I eventually tracked him down he said, 'The harvest is finished and the
rabbis fly out tomorrow, but I can call you back next year.'

$\sim$

You can trust Antonio Miceli to do what he says, and a year later he
called me back to say that the harvest would begin in a week. About

11,000 people live in Scalea for most of the year but by the time the Lubavitchers arrive, in August, the population is swollen by holiday-makers and all the hotels on the seafront are fully booked. The modern part of town is close to the sea and in August the shop-fronts were covered by a psychedelic array of beach toys, an ark's worth of inflatable animals poised for launching. The voices of tourists in the crowded streets were all Neapolitan or Sicilian; Mass was celebrated in the open air; there were ripe figs on the trees near the war memorial; piles of stinking rubbish by the underpass and a Brazilian circus encamped on the edge of town. In winter a river runs through the middle of Scalea, but by August it was dry; no more dark places, no hidden depths, a fluvial subconscious laid bare.

The houses in the medieval part of town are stacked up across the hill, at a safe distance from a sea that brought their inhabitants nothing but trouble. The *Calabresi* lived in dread of invasion from Turkey or North Africa. They were never great fishermen, and local people have told me that pork has always featured large in their diet because pigs were the only thing the invaders wouldn't steal. The medieval town is enclosed by walls, and inside Porta Marittima, the sea gate, the narrow streets are usually deserted. In mid-August, the only sounds came from indoors, where families framed against the dusky interiors of their houses sat over meals that seemed always to be in progress. When darkness fell, a breeze began to blow, the air was cool, the night quiet. Until 2 a.m. Because when Mass was over they moved the altar out of the piazza and made space for the open-air disco, and when that finished, the firework display got started, and in the silence that eventually followed it, cats began their amorous conversations. August nights on the Riviera dei Cedri are too good for sleeping.

To reach the *esrog* groves you must turn your back on Scalea and the sea, and drive in bright sunshine down the narrow, rutted lanes towards the mountains. This inland area is generally much quieter than the Riviera towns, and many of the rabbis and citron merchants wisely remove themselves from the chaos by renting apartments or rooms among the groves. They bring almost all of their own food with them in their suitcases, relying only on a local baker to make their bread under strict supervision.

Citron merchants demand perfection from Calabria's citron farmers because only perfect fruit can be taken into the synagogue during Sukkoth. A perfect *esrog* is the fruit of an ungrafted tree and this poses practical problems for the citron farmer. Any citrus tree grown from seed is less sturdy than a grafted one, and for the tender citron, being grafted on to resilient sour orange rootstock is particularly beneficial. Seed-grown citrons are slow to crop and will only fruit well for ten to twelve years. After that, productivity diminishes each year, and after twenty years it may fail completely.

A perfect *esrog* also has skin that is intact, unblemished by scars or marks caused by insects, and its colour is entirely even. For the farmer, this means visiting the citron grove every few days and removing any thorns or twigs that might cross the path of the developing fruit and damage its skin, or leaves that might shade it and make its colour mottled. The trees must be drenched with expensive insecticide – or *medicina* as it's misleadingly called in Italian – from June, when the fruit begins to develop, until August, when it's harvested. And if the insecticide is too concentrated it leaves white scorch marks on the fruit, rendering it imperfect.

Finally, the fruit must be 'complete'. This means its skin has no holes in it and the style (or *pitom* in Hebrew) of the flower from which it grew will still be attached. This last requirement leads, in the words of one Jewish-American website, to 'esrog-handling heartache'. No wonder, for as the fruit ripens, the *pitom* dries out and becomes ever more fragile. If the *pitom* falls off while the fruit is still on the tree that's all right and the *esrog* is still acceptable. However, if it falls off after the fruit is picked, the *esrog* is considered incomplete. A citron merchant only has to look at the scar left by the *pitom* to know whether it has fallen off before or after the fruit was picked.

There are many people on the Riviera dei Cedri who have worked during the *esrog* harvest ever since they were children. One of them told me that when he was a child, he and his friends decided to put the rabbis and citron merchants to the test. They deliberately presented them with fruit that was almost invisibly scarred by tiny bruises or the minuscule holes made by insects. Naturally the rabbis rejected them, but the children brought the same fruit back again and again. Eventually the rabbis gathered up all the imperfect fruit and threw them into the middle of the river, where the children couldn't reach them.

Miceli introduced me to the Donato, a family who live on the edge of Santa Maria del Cedro. They have been associated with the Kellers, Lubavitcher citron merchants from New York, since the 1970s. At first it was only Rabbi Keller who came to Calabria to buy *esrogim*, but then he began to bring his son Shmuel. When Shmuel first came to Calabria he was fourteen. He wasn't Rabbi Keller's eldest son and yet he showed more interest than his brothers in the strange business of harvesting and selling *esrogim*. When the rabbi died, Shmuel took over the business. He buys fruit for Lubavitcher communities in Puerto Rico, Venezuela, Israel, all over America and in Manchester, England, where an *esrog* can fetch anything from £25 to £250. Any remaining fruit is sold in his shop in New York.

Shmuel is a tall man in his early forties whose views on the world have the absolute certainty born of religious conviction. As a child, he was educated in Jewish schools, followed by a Jewish university in New York. For most of the year he works as a librarian in a Jewish university library. He slips easily into Hebrew in the company of other merchants, and even when he swaps his dark suit and white shirt for jeans and a T-shirt in the *esrog* groves, he's easily identified by his full beard. He is utterly at home in the Calabrian countryside. He has been visiting the Donato family in Santa Maria del Cedro every summer for twenty-seven years, and he always celebrates his birthday with one of their sons. He speaks his own unusual form of Italian, rich in dialect words and the specialized technical terms associated with the intricacies of citron cultivation.

To Pietro and his wife, Shmuel is both an important client and a son. They understand his business and respect the traditions behind it, but they aren't slow to tease or correct him. Sometimes journalists from the local paper turn up to interview or photograph him. Shmuel puts on a suit for these occasions and I once saw Signora Donato dart out of the house as the photographer turned the camera on him. 'Samuele!' she yelled. 'Your shirt!' Then she placed herself squarely between him and the camera. Shmuel towered over her tiny figure as she began to yank up his trousers and tuck in his shirt as if he were still the boy who arrived in Santa Maria del Cedro with his father for the first time, all those years ago.

~

Growing citrons for the *esrog* harvest is fraught with anxiety. On average, only 20 per cent of the crop will be of a high enough quality to pick. Each fruit is carefully examined before picking, but sometimes a fruit's flaws are only revealed after it is removed from the tree. The merchant will pay only for *esrogim* that he deems perfect, so that generally only about 85 per cent of the *esrogim* picked will be paid for. Shmuel and all the other merchants pay the farmers individually for each *esrog*, giving them the same price, regardless of the fruit's size or overall quality. When they get home they divide the fruit into three categories, allotting each one a different price. Shmuel always buys a little bit of damaged fruit that he can sell off for next to nothing to teachers who might want to pass it around a classroom. A damaged *esrog* might fetch $5 or so in New York, but a perfect fruit of a good size could be worth over $200. The unpicked fruit is left on the trees until November, when a second, very different harvest takes place.

The *esrog* harvest began early, when the air in the citron grove was still cool and the hills and mountains beyond it were swathed in mist. Pietro Donato worked with his two sons, Gianluca and Giuseppe. The atmosphere was tense because it is always difficult to find enough fruit of a high standard, and Shmuel was often driven to smoking. 'I don't smoke at all in America,' he said, 'only here. It's the stress.' Before long we were joined by another citron merchant, this time a Russian. I moved forward to greet him, but he held up his hands. 'I'm sorry,' he said. 'I can't shake your hand because you are a woman.' He wandered away with Shmuel, adding Hebrew to the cocktail of languages being spoken among the trees. As I listened to them, I was struck again by the extraordinary power of the citron to influence human behaviour. What other fruit could carry a cargo of symbolism sufficient to attract this strange combination of a Russian and an American, two Britons, a journalist from Reggio and three Calabrians to a remote field at the bottom of Italy?

Pietro, Gianluca and Giuseppe cast themselves on to their backs beneath the trees and pulled themselves along on their elbows, looking fixedly at the fruit above their heads. 'I'm crawling about down here like a snake,' said Pietro. 'Like a worm,' said Gianluca from the next row. 'And if you make a mistake,' Pietro said resentfully, 'the trees prick you. And you have to pull incredibly hard to get one of those thorns out of your knee, and the pain is almost unbelievable.'

Picking *esrogim* is a slow and meditative business. Pietro lay on his side in the damp shade beneath the trees, complaining gently. He paused beneath a handsome fruit, cupped it gently with one hand and subjected it to slow, intimate scrutiny. He looked like Adam in the Garden of Eden, on the verge of making a big mistake, though a citron grove is no Eden, and the fruit didn't pass the test. He moved on, paused again and reached up to grasp another fruit. This time he was uncertain and he wanted confirmation from Shmuel before he cut the *esrog* from the tree. 'Come and look at this, Samuele!' he shouted. No response. He shouted again, more loudly. 'Don't worry. It's all in God's hands,' Shmuel shouted back. 'No, it's not,' I heard Pietro muttering indignantly. 'It's in my hand.' The trees screened Gianluca's and Giuseppe's upper bodies. All I could see was the slow progress of their legs moving horizontally along the ground like a cartoon strip. Pietro stood for a moment beside me. 'Looking up at the fruit hurts your neck,' he said, 'and you have to look into the sun, and that gives you a headache.'

The worst thing of all, everyone agreed, is the insecticide. The men work in a miasma of smoke from their own cigarettes mixed with insecticide, for the *esrog*, that symbol of purity, is doused in it from the moment the flowers form in spring until the fruit is picked in August. Pietro has been working among citrons and picking their fruit ever since he was a child. The spray rubs off the fruit on to his hands as it's picked. 'Sometimes I lick my finger to rub a mark off the skin, and then I lick it again to clean the next fruit. By the end of the day my lips are black with poison,' Pietro declared. 'Who knows how many litres of the stuff I've absorbed.' I wonder what a perfect *esrog* looked like before insecticide was invented. Perfection in those days must have been a very different thing.

Esrogim used to be picked into baskets lined with linen and Shmuel told me he was the first to import the custom-made foam trays already used during the *esrog* harvest in Israel. The foam has citron-shaped holes cut in it, and the trays can be used to carry each precious cargo of fruit from the trees to a table set up in the shade at the edge of the field. They protect the fruit so effectively from bruising and scratching that everyone uses them now. 'Twenty-five years ago,' Shmuel remembered, 'we used to sit on the ground to check the fruit, but now they know we need a table to work at and a bowl to wash it in.' He lined a number of *esrogim* up in front of him, being careful not to allow the fruit to touch, for fear of bruising. The process of examining and cleaning the fruit is sacred and it is done with scrupulous care. 'This is a pretty fruit,' Shmuel said, 'but it's spoilt by this leaf mark.' I couldn't see it, but he explained that there was a slight difference in the colour of a small area of the skin where it had been shaded from the sun by a leaf. Another fine fruit was irrevocably spoilt by a tiny black blemish. Shmuel diagnosed this as scratching caused by somebody trying to rub a leaf mark off the skin. 'The problem here,' he explained, 'is that a tiny sliver of skin is missing. If it was just the mark it might be acceptable.'

If an *esrog* appeared to be perfect, he lowered it into a plastic washing-up bowl full of water, gently, fondly, like a father lowering his baby into the bath. When it was washed he peered at it again, sometimes using a magnifying glass, took a damp cotton bud and cleaned the deeper crevices in its skin and the awkward area around the stalk, and then he dried it off with a paper towel. All of this had to be done with the greatest care, because it's all too easy to knock off the *pitom* at this stage.

When the fruit was clean and dry Shmuel put each one into its own plastic bag, making a hole in one corner so the *pitom* could stick out without being damaged. Sealing the fruit in a bag flushes out any insects that might, by some miracle, have survived the spraying, and after a week or so Shmuel checks each bag and shakes the insects out of it. In this way he can guarantee that the fruit is insect-free when it is imported into America. Insect-free, but poisonous all the same. When the festival is over, it's traditional to make a marmalade from *esrogim*. Shmuel's wife boils their fruit in seven changes of water in an attempt to remove the insecticide before she makes her marmalade. When the fruit arrives in New York, Shmuel unpacks it, sorts it and sends it off to his clients. His business is international and some of the orders are from European customers. Their fruit has to be flown straight back across the Atlantic. And while all this is happening Pietro Donato and his sons in Calabria straighten their backs for the first time in weeks, then extract the last few citron thorns from their fingers and knees.

～

The ancient Greeks thought the citron inedible, but over the centuries the *Calabresi* have invented numerous uses for the Diamante citrons gathered during the November harvest. The fresh fruit is made into jam, citron squash and a dense and powerful citron liqueur called *cedro* or *cedrello*. In summer citrons are often used to make this startlingly refreshing, albeit somewhat astringent, salad using only the *albedo*, the thick white pith that is its sweetest part.

INSALATA DI CEDRO

1 small or ½ large, ripe citron
1 shallot, sliced lengthways into slivers
100ml olive oil
100ml fresh lemon juice
sea salt and black pepper
a handful of parsley
black olives, stoned

Peel the citron with a sharp potato peeler, being careful to remove all of the bright yellow rind, which is the bitterest part of the fruit. Cut the peeled citron into thick segments. Trim away the central pulp and discard it, so that you are left with only the sweet white pith. Slice each segment of pith into very thin strips and put the strips in a bowl. Add the sliced shallot. Mix the olive oil and lemon juice, pour over the salad and toss together. Season with sea salt and black pepper. Allow the salad to rest for an hour so that the dressing is completely absorbed. Finely chop the parsley and olives before adding them to the salad. Serve immediately.

When I told the *padrona* of La Rondinella restaurant in Scalea that I had come to the Riviera dei Cedri for the *esrog* harvest she immediately filled my glass with *cedrello* and rushed off to the *pasticceria* next door to get something very special to go with it. She returned carrying a small, perfectly packed parcel of leaves neatly tied with a cord made from a dried rush. I untied it gingerly and folded back the leaves. They were citron leaves of course, and the parcel had been baked crisp in the oven. Nestling inside were three or four plump raisins soaked in *cedrello*. Gabriele d'Annunzio, writer and controversial political leader, wrote about these *panicelli*, which have been made for hundreds of years in Calabria. In *Leda without Swan* (1916) he compared the small, rectangular envelopes to 'books sealed up by someone who had mistaken the orchard for a library'. He can have experienced none of my anxiety about spoiling the beauty of the neatly wrapped parcel as he describes breaking through the layers of crisp leaves with his nails until he eventually reached 'the final leaf in which the secret is enfolded, perfumed like bergamot . . . A flavour that delights even before we taste it'. And so it did.

~

November found me back in Scalea for the second citron harvest, a very different occasion from the first. I booked into an empty hotel on the seafront and woke in the morning to the sound of torrential rain. The television in the dining room showed snow in Lombardy, high water in Venice and floods in Tuscany, but there was no mention of the rain that would surely bring the citron harvest in Calabria to a grinding halt.

After breakfast Miceli's daughter, Sara, picked me up in her tiny Smart car, its fragile frame making heavy weather of the flooded and rutted roads. Soon we were deep in Miceli country, passing Antonio Miceli's citron groves, his brother's butchery business and his wife's bar before descending steeply to the jewel in the crown, Miceli's agricultural supplies, greengrocer, ironmonger and garden centre. You might go there to buy a part for your tractor, a string of chilli peppers and a votive candle, or a tin of paint and a dog kennel, a chain saw, a bag of fresh chestnuts and a palm tree, and so many people do that the road to this emporium is always busy. I noticed Sara was wearing a pair of soft leather boots that finished somewhere high above her knees. My own boots were a little sturdier, but neither of us was properly dressed for a downpour in a Calabrian citron grove. She took me to Miceli's wellington boot department and told me to help myself. I could have had white ones, like a vet or the dispatcher at a slaughterhouse, or pink ones with flowers on them, but Sara had other ideas. She found me a pair in sober khaki green that she teamed up with a hunter's camouflage cap, tearing off the price tag before I crammed it on to my head.

Our first stop was a citron grove only a short distance from the store, but we drove there anyway, the car's tiny wipers failing to keep up with the sheer volume of water sluicing over the windscreen in unbroken sheets. We pulled up behind a blue *ape*, a misleading word in English, but in Italian it means 'bee', or the buzzing three-wheeled truck used by farmers and workmen all over Italy. The back of this one was piled high with iridescent green citrons, stacked like a pyramid of raindrops resisting gravity. The trees themselves grew at the bottom of a muddy bank. We skidded down the slope to them and I soon learned that the ideal citron-growing earth of Calabria turns into an especially sticky kind of mud in wet weather, making our boots a few kilos heavier.

Most of the fruit from the November harvest is candied. In Calabria candied citron peel is used in local pastries and ice cream. Elsewhere you'll find it in *panettone* at Christmas and in Siena's heavily spiced *panforte*. Candying is a long and painstaking process. As soon as the citrons are picked they are rinsed and tipped into barrels full of brine, where they are left to soak for a year, just as *chinotti* are in Liguria. The brine softens their tough skin and makes it a little less bitter, and when they come out they are as wet and slippery as fish. Then they must be soaked in fresh water for a week or so to remove the salt. And just to make the process more labour-intensive, the water needs changing every day. Next the fruit is cut in two and the flesh is filleted from it using a special tool, a spoon with sharpened edges. Smaller citrons are diced at this stage because they are used to make candied peel, but bigger ones are cut in half, filleted and left otherwise intact. Now it's time for candying, but even that takes the best part of a week. The fruit has to be repeatedly boiled and cooled in a carefully calibrated sugar syrup to which water or sugar is added every day. Half a candied citron is known as a *coppa* or goblet. The surface of its undulating, glowing skin is pitted with open pores and smells slightly of disinfectant. You might expect it to be soft and overwhelmingly sweet by the end of the laborious candying process, but even a year of uninterrupted bathing cannot conquer the indomitable citron. It holds its own, remaining crisp and slightly astringent to the end. If you buy one it will come in a plastic tub, looking like a bright green fish out of water and weighing a kilo or more. It's said that some people like to eat a *coppa* at a sitting, but that's hard to imagine.

Whatever time of year it is harvested, the citron's skin contains an abundance of essential oil. Most of it is sold to cosmetics and pharmaceutical companies, although a few years ago the owner of an olive mill near Scalea experimented with pressing citron skins and olives together. He found the process rather laborious because the press had to be thoroughly cleaned before it could be used again to mill olives in the traditional way, but the citron-flavoured oil is delicious eaten with white meat or fish, and now it's even being exported to Germany.

In the 1960s and 1970s citrons could still make you rich and the region produced 70,000 quintals (7 million kilos) of them a year. Since then, tourism has developed along the coast and many farmers have been tempted to sell their land off for development. These days Calabria's citron harvest amounts to only 5,000 quintals or 500,000 kilos a year. Not enough fruit for the goods trains that once trundled up from the Riviera dei Cedri to Milan and northern Italy, their wagons loaded with citrons for candying. Not enough fruit to interest the wholesalers, and they have abandoned the market to local dealers who have established a monopoly that allows them to push prices down as low as they choose.

Nevertheless, some people in Calabria are determined to protect citron cultivation. One of them is Giovanni Fazio, the dapper eighty-one-year-old who made me promise not to repeat his hair-raising stories about local politics. He knew that the climate for cultivating citrons was ideal on the Riviera dei Cedri, and so ten years ago he set out to discover why production was declining so rapidly. He came to the conclusion that citrons are unusually difficult and labour-intensive to cultivate, and he has devoted himself to resolving some of the discomforts and making it a more efficient process.

Fazio explained that the legs of the table-like stands used to support citron trees got in the workers' way during the biannual harvest and during pruning, which must be done four times a year. He also noticed that citron trees were traditionally planted in rows so close together that they could not be accessed with tractors or other modern machinery. This meant that pickers had to lug each heavy box of fruit to the end of the long row of trees before it could be loaded on to a trailer. He invited me to visit his own citron groves, planted ten years ago, to see what he had done to resolve some of these problems.

The following day the sky was clear and the sun warm, a good enough day to harvest citrons. Sara was coming to pick me up, and to pass the time I walked along the empty autumn beach in front of the hotel, where gentle waves drowned out the sound of traffic. I strolled past Torre Talao, part of a sixteenth-century defence system built by the Spanish against Turkish invasion, and among the trees at the top of the beach I spotted the tanned, near-naked figure of an elderly man performing an especially strenuous kind of press-up. He stood for a moment to gaze at me and then fell back to work.

We arrived in Fazio's citron grove just after the lorries, enormous things churning the mud beside the track. The soaking ground had begun to steam in the warmth of the sun, and the primitive stink of stagnant water mixed with the heavy fragrance of the fruit. The rows of trees that emerged from this primeval swamp were generously spaced. They rested on a new kind of support designed by Fazio. It had a single, central leg connected to a technically advanced form of lightweight steel wire to raise the trees' trailing branches. Fazio had experimented with the height of the wires, managing to raise them slightly, making it easier to cultivate around the base of the trees and pick the fruit.

Grass cutting and hoeing around the trees have always been done by hand, but Fazio has designed a mower with a retractable blade that moves in and out as it encounters obstacles, making it ideal for use in the cramped spaces around the trees. He has also raised the height of the netting used to protect citron groves from the threat of frosts, hail storms or other harsh weather in winter. This gives greater movement of air, so that cold air can disperse quickly, leaving the trees undamaged.

Fazio's inventions seem to spell hope for the citron industry and his trees had a managed, modern look. Yet time has done nothing to soften their disposition and the small team of pickers working their way along the rows were equipped for battle. They wore hats, boots and thick jackets and they carried grappling irons. Each picker had an individual harvesting style but all of them were forced to kneel or even lie on the steaming ground to reach their goal. Some thrust the grappling iron between the wet branches, gripped a fruit by its stalk and pulled. Others grasped its bulk between their hands and wrestled long and hard until it came free.

However they managed to wrench the fruit from the tree, they were drenched with water by the soaking branches and their arms, heads, hands and faces were lacerated by the citron's razor-sharp thorns. 'It's like working with fierce animals,' I said to one of the pickers. 'Yes,' she said, glancing at the others. 'They're bad enough, but the trees are even worse.' No one else spoke. The only sounds were of battle, the clamour of rustling leaves and a thud as the heavy fruit hit the ground. A tractor pulling a narrow sledge – another of Fazio's inventions – made its way between the rows. The citrons, most of them muddy, many of them bruised, were flung on to the sledge as it passed, and the whole, casual process looked almost like a deliberate rebellion against the fastidious care that had been taken over the harvest of *esrogim* earlier in the year. And without this contrast, this extraordinary difference in attitude to the same fruit in different contexts, the story of the citron would be incomplete.

The end of the citron harvest also brought an end to the long and fragmented journey I had been making for many years in pursuit of citrus: an end to dodging through traffic in Palermo behind the wheel of tiny hire cars; to getting up early and joining orange pickers outside Catania, or staying up late to celebrate the last night of carnival in Ivrea; to extraordinary citrus-flavoured meals; to biting into sweet lemons and meeting citrus scientists; to observing complex machinery and talking to imaginative entrepreneurs; to tales of speculation, exploitation, disaster and recovery. And yet this is not the end of the story of Italy and its citrus fruit. Citrus has always been a migrant, and although it has moved on to bless other places with wealth from its golden fruit, nothing can rob Italy of its legacy. Its myriad flavours are integral to Italian food, and the scent of *zagara* will always linger on the air of the Mezzogiorno.

The autumn harvest finished with heavy rain, just as it had begun, and in the depths of a sodden citron grove I realized that Italy's love for citrus is mutual. The sight of the trees all around me, roots thrust deep into the drenched ground, leaves shining, fruit glistening like enormous Christmas decorations and branches stretched wide to embrace the unceasing downpour, reminded me of an odd phrase used by Theophrastus in his *Enquiry into Plants*. Citrons will thrive only in a 'soft and well-watered place', he said, '. . . for such places they love'. Water had found its way into the tops of my boots and inside my collar. I was getting more uncomfortable by the minute, and yet all around me the trees exuded a sense of well-being so intense that they seemed actively to love the place they lived, just as Theophrastus said.

Notes

The Scent of Lemons

1 Quoted in Sven Hakon Rossel, *Do You Know the Land Where the Lemon Trees Bloom? Hans Christian Andersen and Italy* (Edizioni Nuova Cultura, 2009), p. 84 (translation by Rossel).
2 Paul Fussell, *Abroad: British Literary Travelling Between the Wars* (Oxford University Press, 1980).
3 Osbert Sitwell, *Discursions on Travel, Art and Life* (Grant Richards, 1925), p. 194.
4 Ibid., p. 196.
5 In D. H. Lawrence, *The Woman Who Rode Away and Other Stories* (Cambridge University Press, 1995), pp. 19–38.
6 Galileo Galilei, *Dialogue Concerning the Two Chief World Systems* (Cambridge University Press, 1953), p. 59.
7 'I Limoni', in Eugenio Montale, *Tutte le poesie* (Mondadori, 1979), p. 17.
8 Joseph Needham, *Science and Civilization in China*, Vol. VI, Part 1 (Cambridge University Press, 1986), p. 104.

Curious Fruit

1 Alex Ramsay and Helena Attlee, *Italian Gardens: A Visitor's Guide* (Robertson McCarta, 1989).
2 Francis Bacon, *Gesta Grayorum* (1594), quoted in Oliver Impey and Arthur MacGregor (eds), *The Origins of Museums and Cabinets of Curiosities in Sixteenth- and Seventeenth-century Europe* (Clarendon Press, 1985), p. 1.

3 Giorgio Vasari, *Lives of the Painters, Sculptors and Architects*, Vol. 2 (Everyman's Library, 1996), pp. 222–3.

4 Cosimo de' Medici, founder of the dynasty, may have introduced this tradition when he combined the family coat of arms – five golden balls and one blue on a gold shield – with the motto *Il pomo d'oro che spunta dal tronco dopo la schiantarsi dal primo, uno avvalso non deficit alter*, or 'The golden apple sprouts from the trunk, and as soon as one is torn away, another takes its place'.

5 Quoted by Giorgio Galletti, 'Agrumi in casa Medici', in Alessandro Tagliolini and Margherita Azzi Visentini (eds), *Il Giardino delle esperidi, gli agrumi nella storia, nella letteratura e nell'arte* (Edifir, 1996), p. 201.

6 Michel de Montaigne, *The Complete Works of Michael de Montaigne*, edited by William Hazlitt (John Templeman, 1845), p. 565.

7 Agostino del Riccio recorded the different varieties of citrus in the Medici collection in a treatise called *Agricoltura sperimentale*, which he began to write in 1585, two years before Francesco I's death.

8 Del Riccio also mentions the custom of picking one particular fruit in the collection, a large, smooth-skinned citron–lemon hybrid called the *Pomo d'Adamo*, or Adam's Apple, and giving it away as a token of friendship.

9 See M. Mattolini, *Il Principe illuminato* (Edizioni Medicea, 1981).

10 Anna Maerker, 'Uses and Publics of the Anatomical Model Collection of La Specola, Florence, and the Josephinium, Vienna, around 1800', in M. Beretta (ed.), *From Private to Public* (Science History Publications, 2005), pp. 81–96.

11 Stefano Casciu, 'Allestimenti di nature morte nelle ville medicee al tempo di Cosimo III: il caso della villa di Topaia', in Stefano Casciu and Chiara Nepi (eds), *Stravaganti e bizzarri, ortaggi e frutti dipinti da Bartolomeo Bimbi per i Medici* (Edifir, 2008), p. 32.

12 Mariachiara Pozzana, *Il Giardino dei frutti, frutteti, orti, pomari nel giardino e nel paesaggio toscano* (Ponte alle Grazie, 1990), pp. 70–72.

13 Another fine collection of wax fruit and vegetables made in the mid-nineteenth century can be seen in Turin at Il Museo della

Frutta Francesco Garnier Valletti (www.museodellafrutta.it/valletti/), although, being in the north of Italy, it does not include citrus fruit.

14 Galeotti is the co-author, with Giorgio and Sergio Tintori, *Ornamental Citrus Plants: Advice on Their Cultivation from Our Rural Gardening Tradition* (Edifir, 2000).

Cooking for the Pope

1 A. Gallo, *Le venti Giornate dell' Agricoltura e de' piaceri della villa* (Bossini, 1775), pp. 185–6.
2 Bartolomeo Scappi, *Opera* (Arnaldo Forni Editore, 1981). Also *The* Opera *of Bartolomeo Scappi (1570): L'arte et prudenza d'un maestro cuoco,* translated with commentary by Terence Scully (University of Toronto Press, 2011).
3 John Dickie, *Delizia! The Epic History of the Italians and Their Food* (Hodder and Stoughton, 2007), pp. 55–6.
4 Samuel Tolkowsky, *Hesperides: A History of the Culture and Use of Citrus Fruits* (J. Bale & Co., 1938), pp. 247–8.
5 Dickie, *Delizia!*, p. 133.
6 Tolkowsky, *Hesperides*, p. 243.

Golden Apples

1 David Mabberley, 'Citrus (Rutaceae): A Review of Recent Advances in Etymology, Systematics and Medical Applications', *Blumea*, Vol. 49 (10 December 2004), p. 484.
2 I am much indebted to Professor David Freedberg for my knowledge of Ferrari's work and his relationship with Dal Pozzo. He has made a thorough study of the correspondence between the two men, and of the 130-page volume of notes about citrus collected by Dal Pozzo on Giovanni Battista Ferrari's behalf, which

is still lodged in the library of the Accademia Nazionale dei Lincei, Via dalle Lungara 10, 00165 Rome.

3 David Freedberg, 'Ferrari on the Classification of Oranges and Lemons', in Elizabeth Cropper, Giovanna Perini and Francesco Solinas (eds), *Documentary Culture: Florence and Rome from Grand Duke Ferdinand I to Pope Alexander VII: Papers from a Colloquium held at Villa Spelman, Florence, 1990* (Nuova Alfa Editorial, 1992), p. 294.

4 Translation by Lily Thompson Hawkinson, taken from *An Introduction to and Notes on the Translation of Hesperides . . ., A Thesis presented to the General Faculty of Claremont College . . .,* 1936.

5 The Swedish naturalist Carl Linnaeus created a binomial system for naming plants that gave each plant a Latin generic name and a specific adjective. Although this modern nomenclature was universally adopted, Linnaean classification of citrus fruit never won universal approval and Ferrari's system continued to be used in Italy long after the eighteenth century.

One of the Sunniest Places in Europe

1 J. W. Goethe, *Italian Journey* (Penguin Books, 1970), p. 226.

2 See John Goodwin, 'Progress of the Two Sicilies Under the Bourbons', *Journal of the Royal Statistical Society*, Vol. V (1842), p. 189.

3 Tyôzaburô Tanaka (1885–1976) is the author of 180 botanical names in the citrus family Rutaceae. His most important publications include *Citrus Fruits of Japan* (1922), *Species Problem in Citrus* (1954) and *Tanaka's Cyclopedia of Edible Plants of the World* (1976). He divided his citrus fruit into 162 species.

4 See Frederick G. Gmitter Jr and Xulan Hu, 'The Possible Role of Yunnan in the Origin of Contemporary *Citrus* Species (Rutaceae)', *Economic Botany*, Vol. 44, No. 2 (1990), pp. 267–77.

5 It was another 300 years before 'orange', derived from the old French *orange*, began to be used to describe colour in Britain.

'Orange' or *arancio* makes a perfectly sensible adjective in English or Italian, where oranges, like the leaves on deciduous trees, turn from green to gold in autumn. These colour changes occur only in the northern hemisphere, where the temperature drops below ten degrees Celsius in autumn, breaking down chlorophyll in the leaves of deciduous trees and the rinds of oranges and triggering the development of the carotenoids that give them both their flaming colours. In his book *Citrus* (University of Chicago Press, 2007), Pierre Laszlo refers to the illogical use of the word 'orange' in Brazil, where it is expressed in Portuguese as *cor-de-laranja*, or 'colour of orange' (pp. 147–8). However, most Brazilian oranges grow in the subtropical climate of São Paulo, where they remain green even when ripe. It might be more effective, Laszlo thinks, to change the adjective to *cor-de-cenoura*, 'colour of carrot'.

6 Xiaomeng Li, Rangjin Xie, Zhenhua Lu and Zhiqin Zhou, 'The Origin of Cultivated Citrus as Inferred from Internal Transcribed Spacer and Chloroplast DNA Sequence and Amplified Fragment Length Polymorphism Fingerprints', *Journal of the American Society for Horticultural Science*, Vol. 135, No. 4 (July 2010), p. 327.

7 Giuseppe Tomasi di Lampedusa, *The Leopard* (Everyman's Library, 1998), p. 132.

8 Andrew M. Watson, *Agricultural Innovation in the Early Islamic World* (Cambridge University Press, 1983), p. 104.

9 Ibid., p. 107.

10 Marisa Positano Distefano, *Agrumi a tavola* (Bonanno Editore, 1995).

11 Abderrahman Tlili, 'La Sicilia descrita dalla penna di un autore del X secolo: Ibn Hawqal' (Biblioteca Virtual Miguel de Cervantes, 2006), pp. 23–32.

12 Ibn Hawqal, *Configuration de la terre (Kibat Surat al Ard)* (Paris, 1964), quoted by Giuseppe Barbera, 'Tra produttività e bellezza: I giardini di agrumi della Conca d'Oro', in Alberta Cazzani (ed.), *Giardini d'agrumi: limoni, cedri e aranci nel paesaggio agrario italiano* (Grafo, 1999), p. 95.

13 Quoted ibid., p. 98.

14 Salvatore Lupo, *Il Giardino degli aranci, il mondo degli agrumi nella storia del Mezzogiorno* (Marsilio Editore, 1990), p. 16.

15 Lois Olson and Helen L. Eddy, 'Ibn-Al-Awam, a Soil Scientist in Moorish Spain', *Geographical Review*, Vol. 33, No. 1 (January 1943), pp. 100–109.

16 Peter Lord, *A Moorish Calendar, from the Book of Agriculture of Ibn al-Awam* (Black Swan Press, 1979).

17 John McPhee, *Oranges* (Farrar, Straus and Giroux, 1967), p. 68.

18 Ibn Jubayr, *The Travels of Ibn Jubayr*, translated by R. J. C. Broadhurst (Jonathan Cape, 1952).

19 From Ibn Jubayr, *Viaggio in Ispagna, Sicilia, Siria, Palestina, Mesopotamia, Arabia, Egitto* (Sellerio Editore, 1979).

20 If you leave the dual carriageway on Via Gafara, it will funnel you straight into Brancaccio and the sprawl of industrial units that surround Castello di Maredolce.

21 David Gilmour, *The Pursuit of Italy: A History of a Land, Its Regions and Their Peoples* (Allen Lane, 2011), p. 88.

22 Leandro Alberti, *Descrittione di tutta Italia* (Vinegia: Presso Altobello Salicato, 1588).

23 Margaret Visser, *Much Depends on Dinner: The Extraordinary History and Mythology, Allure and Obsessions, Perils and Taboos of an Ordinary Meal* (Grove Press, 2010), p. 273.

24 Giuseppe Barbera, *La Conca d'Oro* (Sellerio Editore, 2012), pp. 92–3.

25 Goethe, *Italian Journey*, p. 241.

26 Mignon's famous question from Goethe's *Wilhelm Meister* (1795).

Antiscorbuticks

1 Jeremy Hugh Baron, 'Sailor's Scurvy before and after James Lind – a Reassessment', *Nutrition Reviews*, Vol. 67, No. 6 (June 2009), p. 317.

2   Pierre Laszlo, *Citrus* (University of Chicago Press, 2007), p. 86.
3   Baron, 'Sailor's Scurvy before and after James Lind – a Reassessment', p. 325.

A Golden Bowl of Bitter Lemons

1   Giuseppe Barbera, *Conca d'Oro* (Sellerio Editore, 2012), p. 39.
2   John Dickie, *Cosa Nostra: A History of the Sicilian Mafia* (Coronet, 2004), p. 55.
3   Salvatore Lupo, *Il Giardino degli aranci, il mondo degli agrumi nella storia del Mezzogiorno* (Marsilio Editore, 1990), p. 26.
4   Dickie, *Cosa Nostra*, pp. 26–7.
5   Guy de Maupassant, *La Vie Errante, Allouma, Toine and Other Stories* (Kessinger, 2004), p. 56.
6   Dickie, *Cosa Nostra*, p. 43.
7   John Dickie, *Blood Brotherhoods: The Rise of the Italian Mafias* (Sceptre, 2011), p. 85.
8   Dickie, *Cosa Nostra*, p. 26.
9   Lupo, *Il Giardino degli aranci, il mondo degli agrumi nella*, p. 118.
10 Ibid.
11 Giuseppe Barbera, *Tutti Frutti* (Mondadori, 2007), pp. 80–81.
12 L. Franchetti, *La Sicilia nel 1876. Condizioni politiche e amministrative della Sicilia*, Vol. I of L. Franchetti and S. Sonnino, *Inchiesta in Sicilia* (Florence, 1974).
13 Francesco Calabrese and Vincenzo Vacante, *Citrus, trattato di agrumicoltura* (Edagricole, 2009), p. 57.
14 Ibid., p. 103.
15 Barbera, *Tutti Frutti*, p. 88.
16 Dickie, *Cosa Nostra*, pp. 255–6.
17 Ibid., pp. 356–8.
18 Giuseppe Barbera, 'Conca d'Oro addio. Così Palermo ha perso il tesoro verde', *La Reppublica*, 22 March 2012.

Oranges Soaked in Sunsets

1 Guy de Maupassant, *La Vie Errante, Allouma, Toine and Other Stories* (Kessinger, 2004), p. 64.

2 *Meraviglie d'Italia* ('The Marvels of Italy'), a collection of articles about Gadda's travels published in 1939.

3 Eugenio Buttelli, Concetta Licciardello, Yang Zhang, Jianjun Liu, Steve Mackay, Paul Bailey, Giuseppe Reforgiato-Recupero and Cathie Martin, 'Retrotransposons Control Fruit-Specific, Cold-Dependent Accumulation of Anthocyanins in Blood Oranges', *Plant Cell*, Vol. 24, No. 3 (March 2012), pp. 1242–55.

4 Francesco P. Bonina, Carmelo Puglia, Giuseppina Frasca, Francesco Cimino, Domenico Trombetta, Giovanni Tringali, Annamaria Roccazzello, Elio Insiriello, Paolo Rapisarda and Antonella Saija, 'Protective Effects of a Standardised Red Orange Extract on Air Pollution-Induced Oxidative Damage in Traffic Police Officers', *Natural Product Research*, Vol. 22, No. 17 (20 November 2008), pp. 1544–51.

5 L. Titta, M. Trinei, M. Stendardo, I. Berniakovich, K. Petroni, C. Tonelli, P. Riso, M. Porrini, S. Minucci, P. G. Pelicci, P. Rapisarda, G. Reforgiato Recupero and M. Giorgio, 'Blood Orange Juice Inhibits Fat Accumulation in Mice', *International Journal of Obesity*, Vol. 34, No. 3 (2010), pp. 578–88.

6 Bonina et al., 'Protective Effects of a Standardised Red Orange Extract on Air Pollution-Induced Oxidative Damage in Traffic Police Officers'.

7 Buttelli et al., 'Retrotransposons Control Fruit-Specific Cold-Dependent Accumulation of Anthocyanins in Blood Oranges'.

8 Translated by D. H. Lawrence in *Little Novels of Sicily* (Thomas Seltzer, 1925), p. 68.

9 In the UK, for more certain success you could sign up to Abel and Cole's organic box scheme, because they import oranges directly from Il Biviere. See www.abelandcole.co.uk.

10 Andrew Marvell, *The Poems of Andrew Marvell* (Lawrence & Bullen, 1892), pp. 39–40.

The Runt of the Litter

1 Peter Lord, *A Moorish Calendar, from the Book of Agriculture of Ibn al-Awam* (Black Swan Press, 1979).
2 Giorgio Gallesio, *Orange Culture: A Treatise on the Citrus Family* (The Florida Agriculturalist, Jacksonville, 1876), p. 43 (page number refers to print-on-demand facsimile supplied by General Books).
3 Annalisa Maniglio Calcagno, 'Il Giardino di agrumi in Liguria', in Alessandro Tagliolini and Margherita Azzi Visentini (eds), *Il Giardino degli esperidi, gli agrumi nella storia, nella letteratura e nell'arte* (Edifir, 1996), p. 219.
4 Christiane Garnero Morena, 'L'origine e le vicende del paesaggio agrumicolo della Provenza orientale e della Liguria', in Alberta Cazzani (ed.), *Giardini d'agrumi: I limoni, cedri e aranci nel paesaggio agrario italiano* (Grafo, 1999), p. 77.
5 John Evelyn, *The Diary of John Evelyn* (Everyman's Library, 2006), p. 92.
6 Charles Dickens, *Picture from Italy* (Bradbury and Evans, 1846), p. 54.
7 Ibid., p. 36.
8 Ibid., pp. 51–2.
9 Evelyn, *The Diary of John Evelyn*, pp. 94–5.
10 'I Limoni', in Eugenio Montale, *Tutte le poesie* (Mondadori, 1979), p. 18.
11 Arthur Young, *Travels in France and Italy during the Years 1787, 1788 and 1789* (W. Richardson, 1794), p. 202.
12 Quoted by Alistair Moore in *La Mortola: In the footsteps of Thomas Hanbury* (Cadogan, 2004), p. 150.
13 Tobias Smollett, *Travels Through France and Italy* (R. Baldwin, 1767), p. 224.
14 Francesco Calabrese, *La Favolosa storia degli agrumi* (L'Epos, 2004), p. 184.
15 Morena, 'L'origine e le vicende del paesaggio agrumicolo della Provenza orientale e della Liguria', p. 80.

16 M. de la Quintinye, *Instruction pour les jardins fruitiers et potagers, avec un traité des orangers, suivy de quelque réflexions sur l'agriculture* (Henri Desbordes, Amsterdam, 1692).

17 Smollett, *Travels Through France and Italy*, p. 314.

18 Silvia Alborno (ed.), *Monet a Bordighera* (Leonardo International, 1998), p. 81.

19 J. Henry Bennett, *Mentone and the Riviera as a Winter Climate* (Churchill, 1861).

20 Giovanni Ruffini, *Doctor Antonio: A Tale of Italy* (Tauchnitz, 1861), p. 100.

21 Ibid., p. 179.

22 Ibid., p. 309.

23 Quoted by Charles Quest Ritson in *The English Garden Abroad* (Viking, 1992), p. 18.

The Sweet Scent of Zagara

1 Giuseppe Tomasi di Lampedusa, *The Leopard* (Everyman's Library, 1998), p. 8.

2 Ibid., p. 18.

Dogged Madness

1 J. W. Goethe, *Italian Journey* (Penguin Books, 1970), pp. 42–3.

2 Domenico Fava, *I Limoni a Limone sul Garda* (Editrice Cassa Rurale ed Artigiana di Vesio Tremosine, 1985).

3 Leila Losi, *I Giardini dei limoni del Lago di Garda, dal passato al presente* (Druck, 2005), p. 72.

4 D. H. Lawrence, 'The Lemon Gardens', in *Twilight in Italy* (Penguin Books, 1976), pp. 39–61.

Green Gold

1　Edward Lear, *Edward Lear in Southern Italy: Journals of a Land-scape Painter in Southern Calabria and the Kingdom of Naples* (William Kimber and Co. Ltd, 1964), p. 41.

2　John Dickie, *Blood Brotherhoods: The Rise of the Italian Mafias* (Sceptre, 2011), p. 181.

3　Henry Swinburne, *Travels in the Two Sicilies in the years 1777, 1778, 1779*, Vol. II (J. Nichols, 1790).

4　Markus Eckstein, *Eau de Cologne: Farina's 300th Anniversary* (J. P. Bachem Verlag), p. 9.

5　Ibid., p. 4.

6　Lear, *Edward Lear in Southern Italy*, pp. 31–2.

7　Raleigh Trevelyan, *Princes Under the Volcano* (Macmillan, 1972), p. 5.

8　Alfredo Focà, *Sull'azione antimicrobica dell'essenza di bergamotto* (Università degli Studi di Catanzaro, 2001), p. 14.

9　Alfredo Focà, *Dell'essenza di bergamotto* (Franco Pancallo Editore, 2005), pp. 113–25.

10　Giorgio Gallesio, *Orange Culture: A Treatise on the Citrus Family* (The Florida Agriculturalist, Jacksonville, 1876), p. 39 (page number refers to print-on-demand facsimile supplied by General Books).

Unique Harvest

1　Gil Marks, *The Encyclopedia of Jewish Food* (John Wiley & Sons, 2010), p. 178.

2　Samuel Tolkowsky, *Hesperides: A History of the Culture and Use of Citrus Fruits* (J. Bale & Co., 1938).

3　Robin Lane Fox, *Alexander the Great* (Penguin Books, 1987), p. 102.

4　A. G. Morton, *The History of Botanical Science* (Academic Press, 1981), p. 29.

5   Quoted in Tolkowsky, *Hesperides*, p. 91.

6   Sally Grainger and Christopher Grocock, *Apicius: A Critical Edition* (Prospect Books, 2006).

7   Tolkowsky, *Hesperides*, p. 71.

8   Athanaeus, *Deipnosophistae*, Book III.

9   Alfred C. Andrews, 'Acclimatization of Citrus Fruits in the Mediterranean Region', *Agricultural History*, Vol. 35, No. 1 (January 1961), p. 38.

10 Erich Isaac, 'Influence of Religion on the Spread of Citrus', *Science*, Vol. 129, No. 3343 (25 January 1959), p. 180.

11 Ibid., p. 184.

12 Elisabetta Nicolosi, Stefano La Malfa, Mohamed El-Otmani, Moshe Negbi and Eliezer E. Goldschmidt, 'The Search for the Authentic Citron (*Citrus medica* L.): Historic and Genetic Analysis', *HortScience*, Vol. 40, No. 7 (2005), p. 1964.

13 Erich Isaac, 'The Citron in the Mediterranean: A Study in Religious Influences', *Economic Geography*, Vol. 35, No. 1 (January 1959), p. 73.

Places to Visit

Gardens with citrus collections

Many historic gardens in Italy have ornamental citrus collections. Information about visiting those mentioned in the book is given below:

Il Giardino del Biviere, Contrada Biviere, 96016 Lentini, Siracusa, Sicily. A private garden surrounded by fields of organic vegetable and citrus groves. It contains a magnificent collection of palms and succulents built up over many years. Can be visited by appointment only.
See www.ilgiardinodelbiviere.it for contact details and Helena Attlee, *Italy's Private Gardens: An Inside View* (Frances Lincoln, 2010) for further information.

Orto Botanico Florence, Via P. A. Micheli 3, 50121 Florence. The citrus collection in Florence's botanic garden includes a specimen of La Bizzarria, propagated by Paolo Galeotti.
See www.museumsinflorence.com/musei/Botanical_garden.html; tel. (39) 055 2757402.

Villa del Principe, Piazza del Principe 4, 16126 Genoa. The restoration of the garden at Villa del Principe has been realized by Ada Segre. Pot-grown citrus are among many plants once grown here in the sixteenth century and reintroduced after restoration.
See www.dopart.it/genova; email info@palazzodelprincipe.it; tel. (39) 010 255509.

Villa La Pietra, Via Bolognese 120, 50139 Florence.

You can see fine examples of stocky, traditionally grown citrus trees in the walled orchard, or *pomario*, behind Villa La Pietra, many of them several hundred years old. Visits by appointment only.

See www.nyu.edu/global/lapietra (email villa.lapietra@nyu.edu) for contact details, and Helena Attlee, *Italy's Private Gardens: An Inside View* (Frances Lincoln, 2010) for further information.

Villa Medici di Castello, Via di Castello 47, 50141 Florence, and Boboli Gardens, Piazza Pitti, Florence.
The Medici citrus collection is still housed in these two gardens. Both are open to the public.
See www.polomuseale.firenze.it; tel. (39) 055 2388786.

Villa Palagonia, Piazza Garibaldi 3, Bagheria, Palermo, Sicily.
The villa's strange garden incorporates a grove of lemon, orange and grapefruit trees. Open to the public.
See www.villapalagonia.it; tel. (39) 091 932088.

Villa Poggio Torselli, Via Scopeti 10, 50026 San Casciano in Val di Pesa, Tuscany.
The garden was laid out at the beginning of the eighteenth century and has recently been restored under the direction of Ada Segre. There is an extensive citrus collection and a magnificent *limonaia* flanks the site. Visits by appointment only.
See www.poggiotorselli.it; tel. (39) 055 8290241/8229557.

A handful of lemon gardens on Lake Garda have been restored and are open to visitors. For example, Pra de la Fam at Tignale, on the main road along the west coast of the lake, is open every day from 25 April to 27 October, from 10 a.m. to 5.30 p.m.
See www.comune.limonesulgarda.

Castello di Racconigi, Via Morosini 3, 12035 Racconigi, Cuneo.
There is a large citrus collection at Racconigi, but it is the vast *limonaia* that really merits a visit.

See www.ilcastellodiracconigi.it/ita/index.htm; email sbap-to. racconigi@beniculturali.it; tel. (39) 017 284005.

Citrus nurseries

Hortus Hesperidis, Piazza Umberto 1/19, 98056 Mazzarrà S. Andrea, Messina, Sicily.
Giuseppe Messina's magnificent nursery contains several bizarre and unusual citrus varieties.
See www.hortushesperidis.com; tel. (39) 094 183245; mobile (39) 339 2437968.

Vivai Oscar Tintori, Via del Tiro a Segno 55, 51012 Castellare di Pescia, Pistoia.
This family business has become one of the best-known citrus nurseries in Europe. Over 200 varieties of citrus from all over the world can be seen growing in ideal conditions in a large ornamental glasshouse known as the Hesperidarium.
See www.oscartintori.il; tel. (39) 057 2429191.

Farms

Two important citrus-growing estates in Sicily have accommodation available to rent:

San Giuliano, 96010 Villasmundo, Melilli, Siracusa, Sicily.
See www.marchesidisangiuliano.it/en; tel. (39) 093 1959022.

San Giorgio, C. da San Giorgio 1, 96016 Lentini, Siracusa, Sicily.
See www.aranjaya.com; tel. (39) 335 1097146.

Festivals

The Battle of the Oranges takes place in Ivrea at the end of the carnival season every year.
See www.storicocarnevaleivrea.it for dates and further details of events.

The citrus gardens of Buggiano Castello, near Pistoia in Tuscany, are open on two consecutive Sundays at the end of April and the beginning of May every year during a festival known as La Campagna Dentro Le Mura.
See www.borgodegliagrumi.it.

Museums

Natural History Museum, Botanical Collection, Via Giorgio La Pira 4, 50121 Florence.
The botanical museum is part of the Museo di Storia Naturale and is open only by appointment. Botanical waxes, including the citrus waxes made for Pietro Leopoldo, are on display here.
See www.msn.unifi.it/mdswitch.html; to make a booking ring (39) 055 2346760 – English spoken.

Museo del Bergamotto, Via Vittorio Veneto 52, 89121 Reggio Calabria.
This is an agricultural museum with a bergamot theme. Displays include examples of the *macchina calabrese*, used to extract essential oil of bergamot, at every stage of its development.
See www.museodelbergamotto.it; email museobergamotto@gmail.com; tel. (39) 338 2721885 or 342 1471594.

Museo del Cedro, Loc. Impresa, Santa Maria del Cedro, 87020 Cosenza.

Displays include every aspect of citron cultivation, citron products and a film about the history of the *esrog*.
Opening hours: July 9.30–12.30, 5–8.30; August 9.30–12.30, 5–10.
Tel. (39) 098 542598.

Museo di Storia Naturale dell'Università di Firenze – Sezione di Zoologia La Specola, Via Romana 17, 50125 Florence.
The large collection of anatomical waxes displayed here was originally made in the same workshop as the citrus waxes and casts.
Closed Mondays.
See www.msn.unifi.it.

Villa Poggio a Caiano.
Bartolomeo Bimbi's portraits of citrus fruit in Tuscany are now housed here. Visits must be booked in advance.
See www.uffizi.firenze.it/musei; tel. (39) 055 877012.

Guided tours

Cooperativa Amalfitana Trasformazione Agrumi.
Email at cata@starnet.it to arrange a guided tour of the lemon groves in Amalfi.

Restaurants and cafés

Il Bergamotto, Piazza San Francesco da Sales 4, Reggio Calabria.
This is Fortunato Marino's *pasticceria* and ice-cream parlour.
Tel. (39) 096 5682555.

Ristorante Il Mulino, Via delle Cartiere 36, Amalfi.
Tel. (39) 089 872223.

Ristorante La Via del Sale, Via San Francesco da Paola 2, 10123 Turin. Tel. (39) 011 888389.

Marmalade

For suppliers of marmalade made at San Giuliano see www.marchesidisangiuliano.it/en/marmalades.html.

A Citrus Chronology

AD 70 Citrus arrives in Italy for the first time in the form of citrons brought by Jews fleeing Jerusalem and settling in Calabria. Citrons are the only species of citrus to grow in Europe until the arrival of sour oranges and lemons in the ninth century.

3rd century AD The shores of southern Calabria and eastern Sicily, once part of Magna Graecia, are annexed by the Roman Empire.

AD 831 Arabs invade Sicily, bringing sour orange (*Citrus aurantium*) and lemon trees with them.

1091 Beginning of the gradual conquest of Sicily by the Normans, who will rule the island until 1266 and continue to cultivate citrus in gardens and as a crop.

12th century *Limone femminello sfusato amalfitano*, the Amalfi lemon, becomes an established crop on the coast of the Bay of Salerno to either side of Amalfi. Crusaders returning from the Near East bring a slightly sweeter orange called *Citrus aurantium* var. *bigaradia* home with them.

1194 A year of rebellion against the feudal overlord in the small northern Italian town of Ivrea that is still commemorated during the Battle of Oranges.

13th century Franciscan friars introduce the first lemons to the shores of Lake Garda.

1266 Norman rule ends in Sicily when Charles I of Anjou invades and kills the king.

1282 Charles I of Anjou is ejected during a violent uprising known as the Sicilian Vespers. Peter of Aragon is proclaimed King of Sicily, marking the beginning of over 400 years of Spanish domination.

1495 Calabria comes under Spanish rule as part of the Kingdom of the Two Sicilies.

1498 Vasco da Gama discovers the sea route around the Cape of Good Hope, bringing sweet oranges from India to Europe for the first time.

1500 A sailor returning from Vietnam introduces the first *chinotto* to Italy through the port of Savona in Liguria.

1537 Cosimo de' Medici takes power in Florence, marking the beginning of 200 years of Medici rule, and commissions a new garden for Castello, which will soon be home to the finest citrus collection in Europe.

17th century The first truly sweet oranges (*Citrus sinensis*) reach Italy from China and the first bergamot trees appear in Calabria as a result of natural cross-pollination.

1644 The famous *bizzarria*, a chimera, is found growing in a garden near Florence.

1646 Giovanni Battista Ferrari publishes *Hesperides, sive, De Malorum aureorum cultura et usu*, a new taxonomy and compendium of knowledge about citrus that includes the first written record of a blood orange.

1699 Bartolomeo Bimbi starts work on paintings of every kind of fruit growing in Tuscany, completing his citrus canvases in 1715.

1708 Giovanni Maria Farina invents a bergamot-based perfume and calls it Eau de Cologne.

1737 Gian Gastone de' Medici dies without issue and his sister Anna Maria is obliged to bequeath all of the Medici possessions, including the citrus collection, to Francis I, Duke of Lorraine.

1747 James Lind undertakes the first controlled trial of lemon juice in the treatment of scurvy.

1749 Linnaeus includes citrus in his new international system of plant nomenclature.

1750 Bergamot begins to be grown on a commercial scale in Calabria.

1753 James Lind publishes an account of his experiment using lemon juice to treat scurvy.

1765 The Medici citrus collection is inherited by Pietro Leopoldo, Duke of Lorraine.

1775 The opening of Florence's natural history museum, La Specola. Among the exhibits are wax and plaster models of the Medici citrus collection.

1796 Napoleon invades northern Italy and gradually takes control of most of the peninsula. By this time lemons are being grown commercially in lemon houses on the shores of Lake Garda.

19th century Bergamot oil begins to be used to give a distinctive aroma and flavour to Earl Grey tea. By the mid-nineteenth century bergamot grows on an industrial scale in Calabria and Francesco Calabrò discovers the antiseptic and healing properties of its essential oil.

1803 The Admiralty appoints Sicily and Malta as sole suppliers of lemon juice to the British Navy all over the world.

1806–15 Sicily is taken under British administration, and this encourages many British merchants to settle there and get involved in the citrus industry.

1807 The first lemons are exported from Sicily to the United States.

1811 Publication of *Traité du Citrus* by Giorgio Gallesio, one of the leading figures in the early-nineteenth-century development of citrus classification.

1821 Mandarins are recorded for the first time on Italian soil in the inventory of the botanic garden of Palermo.

1832 America drops the excise duty in Italian citrus and exports increase dramatically.

1840 Società di Lago di Garda is set up to protect lemon producers on the shores of Lake Garda from unscrupulous middlemen.

1843 The first mandarin glut in Sicily.

1844 Nicola Barillà invents the *macchina calabrese*, a revolutionary machine for extracting the essential oil from bergamot.

1860 Garibaldi lands in Sicily with his army of 1,000 volunteers. His campaign to free the island from Bourbon rule is part of the Risorgimento, or Unification of Italy. The Admiralty abandons its Sicilian lemon juice suppliers and takes out a contract in the West Indies for the supply of Caribbean limes to the entire navy. Nevertheless, citrus cultivation is still the most lucrative agricultural activity in Europe. Mafia-style activity develops in the lemon gardens of the Conca d'Oro, where gummosis, a potentially fatal disease, attacks many of the trees. In Liguria, Augusto Besio opens a factory for candying citrus fruit.

1862 A steamship is used for the first time to export citrus fruit from Sicily to America.

1867 Water is withheld from Femminello lemons, forcing them to fruit for the first time during August.

1876 Leopoldo Franchetti and Sidney Sonnino launch their investigation into society and criminal activity on the Conca d'Oro and elsewhere in Sicily.

1914–18 (First World War) Lemon houses on Lake Garda are abandoned when the lake shore is evacuated, and the wood used to protect the trees in winter is requisitioned for shuttering trenches on the front line. The Medici citrus collection is ousted from the *limonaia* at Castello to make way for wounded soldiers returning from the front.

1917 American chemist James Currie discovers that citric acid can be produced in a laboratory by feeding glucose to *Aspergillus niger* mould.

1932 The Gardesana road is built along the west shore of Lake Garda, bringing tourists to Limone for the first time, and the San Pellegrino drinks company invents Chinotto.

1937 Albert Szent-Györgyi wins the Nobel Prize for identifying Vitamin C as the antiscorbutic ingredient in lemon juice.

1939 'Avana' mandarin trees growing in the groves of Ciaculli and Croceverde Giardina in Sicily mutate and produce valuable new crop of *mandarini tardivi di Ciaculli* for the first time.

1956 Severe frost kills most of the citrus trees in Liguria, and in Calabria the small town of Cipollina is renamed Santa Maria del Cedro, in honour of the citrons growing in the fields surrounding it.

1969 Richard Nixon declares a war on drugs in America and heroin refineries are shut down in France, giving mandarin farming a new significance as refineries are built beneath the Greco family mandarin groves outside Palermo in Sicily.

1986 The Slow Food movement is founded by Carlo Petrini to preserve and promote local foods that will eventually include *chinotti* and *mandarini tardivi di Ciaculli*, and the San Pellegrino drinks company launches a new *chinotto*-based drink called Chinò.

1994–7 *Project Life* restores the mandarin groves around Ciaculli and Croceverde in Sicily, in order to reinvigorate production and transform the land into an agricultural park for the city. In the mid-1990s the Italian government lifted the controls protecting the citrus industry from exports and the market was soon flooded with cheap fruit from abroad.

1999 Genuine Calabrian bergamot oil given a DOP (*Denominazione di Origine Protetta*).

2000 Leoluca Orlando resigns as mayor of Palermo and *Project Life* collapses.

2003 The *chinotto* is chosen as a symbol of the city of Savona.

2004 The *chinotto* nominated for a *presidio*, one of the small bodies set up within the Slow Food movement to protect traditional products in danger of extinction.

Selected Reading

Harold Acton, *The Last Medici* (Sphere Books, 1988)

Alfred C. Andrews, 'Acclimatization of Citrus Fruits in the Mediterranean Region', *Agricultural History*, Vol. 35, No. 1 (January 1961), pp. 35–46

Enrico Baldini, 'The Role of Cassiano dal Pozzo's Paper Museum in Citrus Taxonomy', in David Freedberg and Enrico Baldini, *Citrus Fruit: A Catalogue Raisonné* (Harvey Millar Publishers, 1997)

Giuseppe Barbera, *Tutti Frutti* (Mondadori, 2007)

—, *La Conca d'Oro* (Sellerio Editore, 2012)

—, 'Tra produttività e bellezza: I giardini di agrumi della Conca d'Oro', in Alberta Cazzani (ed.), *Giardini d'agrumi: limoni, cedri e aranci nel paesaggio agrario italiano* (Grafo, 1999)

Jeremy Hugh Baron, 'Sailor's Scurvy before and after James Lind – a Reassessment', *Nutrition Reviews*, Vol. 67, No. 6 (June 2009), pp. 315–32

J. Henry Bennett, *Mentone and the Riviera as a Winter Climate* (Churchill, 1861)

Emanuel Bonavia, *The Cultivated Oranges, Lemons etc. of India and Ceylon* (W. H. Allen & Co., 1888)

Francesco P. Bonina, Carmelo Puglia, Giuseppina Frasca, Francesco Cimino, Domenico Trombetta, Giovanni Tringali, Annamaria Roccazzello, Elio Insiriello, Paolo Rapisarda and Antonella Saija, 'Protective Effects of a Standardised Red Orange Extract on Air Pollution-Induced Oxidative Damage in Traffic Police Officers', *Natural Product Research*, Vol. 22, No. 17 (20 November 2008), pp. 1544–51

Eugenio Butelli, Concetta Licciardello, Yang Zhang, Jianjun Liu, Steve Mackay, Paul Bailey, Giuseppe Reforgiato-Recupero and Cathie Martin, 'Retrotransposons Control Fruit-Specific,

Cold-Dependent Accumulation of Anthocyanins in Blood Oranges', *Plant Cell*, Vol. 24, No. 3 (March 2012), pp. 1242–55

Francesco Calabrese, *La Favolosa storia degli agrumi* (L'Epos, 2004)

Francesco Calabrese and Vincenzo Vacante, *Citrus, trattato di agrumicoltura* (Edagricole, 2009)

Vittorio Caminiti, *Viaggiando mangiando Mangiando viaggiando: Itinerario turistico-gastronomico dal 1847 a tavola con Edward Lear nella provincia di Reggio Calabria* (Provincia di Reggio Calabria, 2001)

Stefano Casciu and Chiara Nepi (eds), *Stravaganti e bizzarri, ortaggi e frutti dipinti da Bartolomeo Bimbi per i Medici* (Edifir, 2008)

Alberta Cazzani (ed.), *Giardini d'agrumi: limoni, cedri e aranci nel paesaggio agrario italiano* (Grafo, 1999)

Horatio Clare, *Sicily: Through Writers' Eyes* (Eland, 2006)

Craig Clunas, *Fruitful Sites: Garden Culture in Ming Dynasty China* (Reaktion Books, 1996)

William C. Cooper, *The Odyssey of the Orange in China* (William C. Cooper, 1990)

Charles Dickens, *Picture from Italy* (Bradbury and Evans, 1846)

John Dickie, *Cosa Nostra: A History of the Sicilian Mafia* (Coronet, 2004)

—, *Delizia! The Epic History of the Italians and Their Food* (Hodder and Stoughton, 2007)

—, *Blood Brotherhoods: The Rise of the Italian Mafias* (Sceptre, 2011)

Marisa Positano Distefano, *Agrumi a tavola* (Bonanno Editore, 1995)

Giovanni Dugo and Angelo di Giacomo, *Citrus: The Genus Citrus* (Taylor and Francis, 2002)

John Evelyn, *The Diary of John Evelyn* (Everyman's Library, 2006)

Domenico Fava, *I Limoni a Limone sul Garda* (Editrice Cassa Rurale ed Artigiana di Vesio Tremosine, 1985)

Giovanni Battista Ferrari, *Hesperides, sive, De Malorum Aureorum cultura et usu, libri quatuor* (1646)

Alfredo Focà, *Sull'azione antimicrobica dell'essenza di bergamotto* (Università degli Studi di Catanzaro, 2001)

—, Focà, *Dell'essenza di bergamotto* (Franco Pancallo Editore, 2005)

Robin Lane Fox, *Alexander the Great* (Penguin, 1987)

David Freedberg, *The Eye of the Lynx: Galileo, His Friends, and the Beginning of Modern Natural History* (University of Chicago Press, 2002)

—, 'From Hebrew and Gardens to Oranges and Lemons: Giovanni Battista Ferrari and Cassiano dal Pozzo', in Francesco Solinas (ed.), *Cassiano dal Pozzo: atti del seminario internazionale di studi* (De Luca, 1989)

—, 'Ferrari on the Classification of Oranges and Lemons', in Elizabeth Cropper, Giovanna Perini and Francesco Solinas (eds), *Documentary Culture: Florence and Rome from Grand-Duke Ferdinand I to Alexander VII: Papers from a Colloquium held at Villa Spelman, Florence, 1990* (Nuova Alfa Editorial, 1992)

—, 'Ferrari and the Pregnant Lemons of Pietrasanta', in Alessandro Tagliolini and Margherita Azzi Visentini (eds), *Il Giardino delle esperidi, gli agrumi nella storia, nella letteratura e nell'arte* (Edifir, 1996)

David Freedberg and Enrico Baldini, *Citrus Fruit: A Catalogue Raisonné* (Harvey Millar Publishers, 1997)

Paul Fussell, *Abroad: British Literary Travelling Between the Wars* (Oxford University Press, 1980)

Galileo Galilei, *Dialogue Concerning the Two Chief World Systems*, translated by Stillman Drake (Cambridge University Press, 1953)

Giorgio Gallesio, *Traité du citrus* (1811); reissued as *Orange Culture: A Treatise on the Citrus Family* (The Florida Agriculturalist, Jacksonville, 1876)

Giorgio Galletti, 'Agrumi in casa Medici', in Alessandro Tagliolini and Margherita Azzi Visentini (eds), *Il Giardino delle esperidi, gli agrumi nella storia, nella letteratura e nell'arte* (Edifir, 1996)

J. H. Galloway, 'The Mediterranean Sugar Industry', *Geographical Review*, Vol. 67, No. 2 (April 1977), pp. 177–94

David Gilmour, *The Pursuit of Italy: A History of a Land, Its Regions and Their Peoples* (Allen Lane, 2011)

J. W. Goethe, *Italian Journey*, translated by W. H. Auden and Elizabeth Mayer (Penguin Books, 1970)

Sally Grainger and Christopher Grocock, *Apicius: A Critical Edition* (Prospect Books, 2006)

Penelope Hobhouse, *Gardens of Persia* (Kales Press, 2004)

Oliver Impey and Arthur MacGregor (eds), *The Origins of Museums and Cabinets of Curiosities in Sixteenth- and Sevententh-century Europe* (Clarendon Press, 1985)

Erich Isaac, 'The Citron in the Mediterranean: A Study in Religious Influences', *Economic Geography*, Vol. 35, No. 1 (January 1959), pp. 71–8

—, 'Influence of Religion on the Spread of Citrus', *Science*, Vol. 129, No. 3343 (23 January 1959), pp. 179–86

Giuseppe Tomasi di Lampedusa, *The Leopard* (Everyman's Library, 1998)

Pierre Laszlo, *Citrus* (University of Chicago Press, 2007)

D. H. Lawrence, *Little Novels of Sicily* (Thomas Seltzer, 1925)

—, 'The Lemon Gardens', in *Twilight in Italy* (Penguin Books, 1976)

—, *The Woman Who Rode Away and Other Stories* (Cambridge University Press, 1995)

—, *Sea and Sardinia* (Cambridge University Press, 1997)

Edward Lear, *Edward Lear in Southern Italy: Journals of a Landscape Painter in Southern Calabria and the Kingdom of Naples*, with an introduction by Peter Quennell (William Kimber and Co. Ltd, 1964)

Peter Lord, *A Moorish Calendar, from the Book of Agriculture of Ibn al-Awam* (Black Swan Press, 1979)

Leila Losi, *I Giardini dei limoni del Lago di Garda, dal passato al presente* (Druck, 2005)

Salvatore Lupo, *Il Giardino degli aranci, il mondo degli agrumi nella storia del Mezzogiorno* (Marsilio Editore, 1990)

David Mabberley, 'Citrus (Rutaceae): A Review of Recent Advances in Etymology, Systematics and Medical Applications', *Blumea*, Vol. 49 (10 December 2004), pp. 481–98

John McPhee, *Oranges* (Farrar, Straus and Giroux, 1967)

Anna Maerker, 'Uses and Publics of the Anatomical Model Collection of La Specola, Florence, and the Josephinum, Vienna, around

1800', in M. Beretta (ed.), *From Private to Public* (Science History Publications, 2005)

Gil Marks, *The Encyclopedia of Jewish Food* (John Wiley & Sons, 2010)

M. Mattolini, *Il Principe illuminato* (Edizioni Medicea, 1981)

Guy de Maupassant, *La Vie Errante, Allouma, Toine and Other Stories* (Kessinger, 2004)

Eugenio Montale, *Tutte le poesie* (Mondadori, 1979)

Alistair Moore, *La Mortola: In the Footsteps of Thomas Hanbury* (Cadogan, 2004)

Christiane Garnero Morena, 'L'origine e le vicende del paesaggio agrumicolo della Provenza orientale e della Liguria', in Alberta Cazzani (ed.), *Giardini d'agrumi: limoni, cedri e aranci nel paesaggio agrario italiano* (Grafo, 1999)

A. G. Morton, *The History of Botanical Science* (Academic Press, 1981)

Joseph Needham, *Science and Civilization in China*, Vol. VI, Part 1 (Cambridge University Press, 1986)

Elisabetta Nicolosi, Stefano La Malfa, Mohamed El-Otmani, Moshe Negbi and Eliezer E. Goldschmidt, 'The Search for the Authentic Citron (*Citrus medica* L.): Historic and Genetic Analysis', *HortScience*, Vol. 40, No. 7 (2005), pp. 1963–8

Lois Olson and Helen L. Eddy, 'Ibn-Al-Awam, a Soil Scientist in Moorish Spain', *Geographical Review*, Vol. 33, No. 1 (January 1943), pp. 100–109

Mariachiara Pozzana, *Il Giardino dei frutti, frutteti, orti, pomari nel giardino e nel paesaggio toscano* (Ponte alle Grazie, 1990)

Franco M. Raimondo and H. Walter Lach, *Le Mele d'Oro: l'affascinante mondo degli agrumi* (Edizioni Grifo-Palermo, 1997)

Gillian Riley, *The Oxford Companion to Italian Food* (Oxford University Press, 2007)

Charles Quest Ritson, *The English Garden Abroad* (Viking, 1992)

Sven Hakon Rossel, *Do You Know the Land Where the Lemon Trees Bloom? Hans Christian Andersen and Italy* (Edizioni Nuova Cultura, 2009)

Giovanni Ruffini, *Doctor Antonio: A Tale of Italy* (Tauchnitz, 1861)

Osbert Sitwell, *Discursions on Art, Travel and Life* (Grant Richards, 1925)

Tobias Smollett, *Travels Through France and Italy* (R. Baldwin, 1767)

Henry Swinburne, *Travels in the Two Sicilies in the years 1777, 1778, 1779,* Vol. II (J. Nichols, 1790)

Alessandro Tagliolini and Margherita Azzi Visentini (eds), *Il Giardino degli esperidi, gli agrumi nella storia, nella letteratura e nell'arte* (Edifir, 1996)

L. Titta, M. Trinei, M. Stendardo, I. Berniakovich, K. Petroni, C. Tonelli, P. Riso, M. Porrini, S. Minucci, P. G. Pelicci, P. Rapisarda, G. Reforgiato Recupero and M. Giorgio, 'Blood Orange Juice Inhibits Fat Accumulation in Mice', *International Journal of Obesity,* Vol. 34, No. 3 (2010), pp. 578–88

Samuel Tolkowsky, *Hesperides: A History of the Culture and Use of Citrus Fruits* (J. Bale & Co., 1938)

Raleigh Trevelyan, *Princes Under the Volcano* (Macmillan, 1972)

Giorgio Vasari, *Lives of the Painters, Sculptors and Architects* (Everyman's Library, 1996)

Margaret Visser, *Much Depends on Dinner: The Extraordinary History and Mythology, Allure and Obsessions, Perils and Taboos of an Ordinary Meal* (Grove Press, 2010)

Andrew M. Watson, *Agricultural Innovation in the Early Islamic World* (Cambridge University Press, 1983)

William Whiston, *The Works of Josephus: Complete and Unabridged* (Hendrickson Pub.; updated edition, 1980)

Arthur Young, *Travels in France and Italy during the Years 1787, 1788* and *1789* (W. Richardson, 1794)

Index